RENEWABLE ENERGY

A PRIMER FOR THE TWENTY-FIRST CENTURY

BRUCE USHER

Columbia University Press *New York*

Columbia University Press
Publishers Since 1893
New York   Chichester, West Sussex
cup.columbia.edu

Library of Congress Cataloging-in-Publication Data
Names: Usher, Bruce, author.
Title: Renewable energy : a primer for the twenty-first century /
Bruce Usher.
Description: New York : Columbia University Press, [2019] |
Series: Columbia University earth institute sustainability primers |
Includes bibliographical references and index.
Identifiers: LCCN 2018024274 | ISBN 9780231187848 (cloth) | ISBN
9780231187855 (pbk.) | ISBN 9780231547529 (e-book) Subjects: LCSH:
Renewable energy sources.
Classification: LCC TJ808 .U84 2019 | DDC 333.79/4—dc23
LC record available at https://lccn.loc.gov/2018024274

Cover design: Julia Kushnirsky
Cover image: © Benjamin Van Der Spek / EyeEm / Getty Images

RENEWABLE ENERGY

COLUMBIA UNIVERSITY

EARTH INSTITUTE SUSTAINABILITY PRIMERS

COLUMBIA UNIVERSITY
EARTH INSTITUTE SUSTAINABILITY PRIMERS

The Earth Institute (EI) at Columbia University is dedicated to innovative research and education to support the emerging field of sustainability. The Columbia University Earth Institute Sustainability Primers series, published in collaboration with Columbia University Press, offers short, solutions-oriented texts for teachers and professionals that open up enlightened conversations and inform important policy debates about how to use natural science, social science, resource management, and economics to solve some of our planet's most pressing concerns, from climate change to food security. The EI primers are brief and provocative, intended to inform and inspire a new, more sustainable generation.

CONTENTS

Preface: Setting the Record Straight *vii*

1 Renewable Energy in the Twenty-First Century 1

2 Energy Transitions: Fire to Electricity 7

3 The Rise of Renewables 21

4 Renewable Wind Energy 29

5 Renewable Solar Energy 43

6 Financing Renewable Energy 55

7 Energy Transitions: Oats to Oil 75

8 The Rise of Electric Vehicles 81

9 Parity 97

10 Convergence 117

11 Consequences 133

12 No Time to Lose 151

Appendix A. Levelized Cost of Electricity (LCOE) *161*

Appendix B. The Transition to Renewable Energy *163*

Glossary *165*

Notes *169*

Index *199*

PREFACE

Setting the Record Straight

A FRIEND of mine recently installed solar panels on her workplace in New York City. This is no small feat. Most buildings in New York are surrounded by even taller buildings which block sunlight for much of the day. But the roof of my friend's building is clear, so she contracted a local company to install the solar panels.

A few months later I had the opportunity to visit, along with several other friends. We all looked at the solar panels on the roof. One friend expressed astonishment that the panels could generate enough electricity to supply the entire building. Another worried that they would be a poor investment. A third friend maintained that solar would always need government subsidies to be financially viable for consumers. They are all smart and highly educated, but they were wrong on every point.

From this experience I began to ask myself: Why do so many people know so little about renewable energy? The rapid growth in renewable wind and solar is not a secret. Towering wind turbines and shimmering solar panels are hard to miss. Yet most people are woefully unaware of what is happening in the energy sector. Bill Ritter, the former governor of Colorado, identified the source of the problem: "People have misconceptions about the cost of renewable energy largely because the public conversation about these resources has been in the form of TV campaign ads and campaign debates, where the truth is, at best, elusive."[1]

Misinformation from politicians has obscured a remarkable energy transition by which renewables have become cost competitive with fossil fuels. And the rise of renewables is only half the story. The transportation sector is also undergoing a transition, from gasoline- and diesel-powered vehicles to electric vehicles. Energy and transportation, two of the world's largest and most valuable industries, are entering a period of dramatic and interlinked change.

As a business school professor, I find these changes interesting. Much of the material in this book stems from the classes I teach at Columbia Business School, but I have more than an academic interest in renewable energy. I am also an investor in early-stage companies. Like many investors, I focus on long-term trends, such as the changes occurring in the energy and transportation sectors. While the timing of the transition to renewable energy and electric vehicles is challenging to predict, the overall trends are clear, creating exciting opportunities for investment.

This brings me to why I have written this book: so that everyone—my friends included—might understand the energy transition to renewables already underway: to set the record straight. Awareness of this trend matters because how we produce and consume energy is critical to our way of life and to the future of our planet. We all need to be making smart decisions about energy. We need our politicians and governments to do so as well. I hope this short book helps in that regard.

• • •

My thanks to Patrick Fitzgerald and Brian Smith of Columbia University Press for embracing a first-time author. Special thanks to Steph Shaw for her exceptional research and editing, and to Michael Gerrard, Dave Kirkpatrick, and Charlie Donovan for taking the time to review and provide feedback. To my brother, Eric Usher, thank you for leading me into this sector in the first place. Naomi, Ben, and Theo, thank you for supporting my career switch to academia. And to my students at Columbia Business School, thank you for motivating and challenging me every day.

RENEWABLE ENERGY

FIGURE 1.1 The Earth. (Source: http://deskbg.com/)

1

RENEWABLE ENERGY IN THE
TWENTY-FIRST CENTURY

Few would deny that the energy transition is underway.

—International Monetary Fund, 2017

R ENEWABLE energy is a surprisingly polarizing subject. Environmentalists are drawn to wind and solar power by the threat of climate change, while economists fret about the cost of replacing coal, natural gas, and oil. Meanwhile, politicians take positions for or against renewables as if taking a moral stand. Despite all the controversy, renewables are already replacing fossil fuels as the world's primary source of energy, a development that is both inevitable and impactful. This book explains that transition and the consequences to follow.

THE RISE OF RENEWABLES

The transition from fossil fuels to renewables in the twenty-first century mirrors previous energy transitions in human history: from wood to coal for heat, and from animal feed to oil for transportation. In each case, the change in energy source was developed using the best technologies of the time, and the transition yielded a better product at a cheaper price. In many countries, highly advanced wind and solar power facilities can now generate electricity at a price that is lower than that of fossil fuels. And competitive pricing is driving consumer demand for renewables, sometimes in the most unlikely places.

In Texas, headquarters to many of America's fossil fuel companies, wind power accounted for 17 percent of the electricity generated in 2017 and is forecast to surpass coal within two years.[1] Great Britain, cradle of the Industrial Revolution, crossed a threshold on April 21, 2017, when the country went a day without burning any coal after four hundred years of constant use.[2] Even Saudi Arabia, owner of the world's most valuable petroleum reserves, is joining the transition, installing several of the world's largest solar projects.[3]

Wind and solar have become the fastest growing sources of new power generation globally, benefitting from a virtuous cycle in which constantly improving technology and declining manufacturing costs result in increased demand and lower prices. Every year, wind and solar become cheaper and more competitive against coal, oil, and natural gas. But that is only the beginning. In the future, renewable energy will continue to increase its market share due to a parallel energy transition occurring in the transportation sector.

TRANSITION IN TRANSPORTATION

One hundred years ago, the gasoline-powered automobile replaced the horse as the primary means of transportation, providing faster, cheaper transport. The transportation sector is now in the early stages of its next great transition, from the internal combustion engine to electric vehicles. Within the next decade, electric vehicles will become less expensive than traditional automobiles, accelerating the transition. This matters to the growth prospects of renewable energy because the emerging market for electric vehicles is rapidly driving down the cost of batteries, and with it the cost of storing electricity.

Renewable wind and solar power are cheap, but they are intermittent, as electricity is only generated in windy or sunny conditions. Fulfilling the potential of a transition from fossil fuels to renewables requires a cost-competitive form of energy storage. Low-cost batteries provide inexpensive storage of electricity generated by wind and solar, enabling consumers to use renewable power even on cloudy or windless days. Inexpensive storage removes the last major hurdle to the next energy transition.

WHY THE ENERGY TRANSITION MATTERS

Energy transitions of the past—from wood to coal to oil and natural gas—were critically important to the development of modern economies. The next energy transition, from fossil fuels to renewables, will be critical to avoiding catastrophic climate change. It will also result in significant changes in geopolitical power: fossil fuel–exporting nations will experience declining revenue while China, India, and many developing countries will benefit from the transition. Global health will improve as coal use and associated particle pollution decreases. And the construction of renewable energy projects will create millions of jobs worldwide.

The growth of renewable energy in the twenty-first century will create winners and losers. Incumbent fossil fuel companies risk losing trillions of dollars in shareholder value; some of them will develop strategies to join the energy transition, while the less nimble will face bankruptcy. Similarly, some countries will lead the transition from fossil fuels and reap long-term benefits, while others will trail behind.

The transition from fossil fuels to renewables has become inevitable as increasingly cheaper wind and solar power replace coal, oil, and natural gas. Governments cannot change the overall trajectory of this energy transition, but their policies do affect the speed of the transition. Government policies and decisions also determine which countries and which players lead the transition and which follow. As in previous energy transitions, there is much at stake.

A PRIMER ON RENEWABLE ENERGY

This book is a primer on the economic fundamentals driving the global transition to renewables. By design it simplifies many points, as the intention is to provide the reader with a solid understanding of the fundamentals, rather than a comprehensive examination of every topic. The focus is primarily on the nexus between the two leading renewable growth sectors—wind and solar—and fossil fuels. Policy issues are surveyed only at a very high level. This book does not advocate for renewable energy from an ethical perspective; it provides a dispassionate examination of the current state of the market, assesses trends, and offers forecasts for the future. In short, this primer provides the reader with a clear-eyed explanation for the extraordinary growth in renewable wind and solar energy to date and the likely trajectory in the years ahead.

This book does engage in speculation on one front: the assessment of consequences. The transition from fossil fuels to renewable energy is a near certainty, but the timing and the resulting geopolitical, economic, health, and climate consequences are less clear. There will be winners—countries, industries, and individuals who will pioneer the transition to renewables and reap the many benefits—and there will be many losers, those who are unable to, or fail to, adapt. But predicting the future first requires a solid understanding of the past.

FIGURE 2.1 Heinrich Füger, *Prometheus Brings Fire to Mankind*, 1817.

2

ENERGY TRANSITIONS

Fire to Electricity

Civilization began with fire. The heat produced by fire became the first form of energy harnessed by humans, providing our earliest ancestors with warmth, light, and cooked sustenance. This phenomenon is captured vividly in the Greek myth of Prometheus, who stole fire from Mount Olympus and bestowed it upon humanity, thus enabling the development of civilization. For this transgression he was punished by Zeus: chained to a rock and tortured for eternity. The gruesome nature of the punishment reflected a recognition of the importance of fire—and thus energy—to the future of humanity.

For thousands of years, humans generated energy by burning wood and other plant biomass. Wood was plentiful in most places, easy to gather and transport, and simple to burn, which made it an attractive and inexpensive source of energy. It seemed to be an endlessly renewable source as well. Throughout the Middle Ages, the vast forests covering much of the planet provided more than enough wood to satisfy humankind's needs for energy production. Historical data from this period is scarce, but the Domesday Book, compiled in England under William the Conqueror in 1086, intimates that the country was 50 percent wooded at the time.[1]

With abundant forests that could easily be harvested and the timber easily transported, wood was the primary source of energy in Britain until the 1600s and in the United States through the 1700s.[2] But population growth, especially in urban centers, led to deforestation when tree removal surpassed the forests' ability to regenerate. The population of England and Wales ballooned from about three million in the early 1530s to nearly double that by 1690.[3] Beyond the proliferation of the use of wood in cooking and heating, population growth also led to increasing demand for lumber for construction and industry. Smelting and ore refining required high levels of energy, which at the time was fueled by trees.[4]

With deforestation came an inevitable economic outcome: the price of wood increased as harvesting moved farther and farther away from major population centers. Denizens of seventeenth-century London, the largest city in the world at the time, felt the rising price of wood in their daily lives. As the price increased, people began to search for a cheaper alternative, ushering in the world's first energy transition.[5]

THE TRANSITION FROM WOOD TO COAL

Coal's advantages were captured by the great American poet Ralph Waldo Emerson. "We may well call it black diamonds. Every basket is power and civilization," he wrote.[6] The benefits of coal over wood make it easy to understand Emerson's enthusiasm. On an equivalent volume basis, coal generates more heat than wood, making it cheaper to gather and transport. This is referred to as energy density, defined as the amount of energy that can be stored per unit volume or mass. On a per-acre basis, a coal field has several thousand times more energy than does a forest.[7]

Coal is also abundant, though it must first be mined from underground and then transported.

While it offered benefits, coal also had drawbacks which impacted quality of life. According to historian Peter Brimblecombe, "Well-bred ladies would not even enter rooms where coal had been burnt . . . and the Renaissance Englishman was not keen to accept beer tainted with the odor of coal smoke."[8] For this and other reasons, the British nobility was resistant to switching from wood to coal. The poor transitioned to coal first, out of necessity as they could no longer afford wood. Nobility began to accept coal after 1603, when Queen Elizabeth died and the new King James used coal in his residence.[9] The first energy transition was under way.

Coal was initially used only for heating homes and was burned in fireplaces in small quantities. The demand for coal began to grow in 1790 when Scottish inventor James Watt introduced a steam engine that could be used in almost any factory, accelerating the Industrial Revolution, first in England and then throughout Europe and across the Atlantic. Watt's steam engine was powered by coal.

In the United States, the transition to coal was even slower, primarily because there was more forested land, making wood correspondingly cheaper. Wood's abundance and proximity to the people who used it meant that there was no economic incentive to search for or develop another source of energy. In 1826, residents of Philadelphia could heat their homes with wood at a lower cost than coal-fueled heating.[10] But coal prices declined as American railroads constructed tracks to access and transport it, falling from $7–$10 per ton in the 1830s to only $3 per ton by the mid-1850s.[11] Low prices encouraged homeowners to switch from wood to coal, and U.S. coal consumption grew more than a hundredfold in the subsequent fifty years.[12]

In the second half of the nineteenth century, coal replaced wood as the primary source of energy in Europe and the United States.[13] Coal mines provided vast amounts of coal to rapidly expanding factories and railways, and the availability of cheap coal was supported by infrastructure decisions made by the government. For example, Great Britain's development of its inland waterway channels made coal 50 percent cheaper to transport. In the frenzy of industrialization, inventors sought to harness the energy from coal for new and more beneficial uses. It was not long before coal's next great use was discovered—the generation of electricity.

COAL-FIRED ELECTRICITY

Electricity was poorly understood and was no more than a curiosity until well into the eighteenth century. Benjamin Franklin conducted his legendary research on electricity by attaching a metal key to the string at the bottom of a kite he then flew in a storm, demonstrating that lightning was electrical in nature (and simultaneously demonstrating that he was a very lucky man; subsequent replications of Franklin's experiment resulted in several deaths).[14] But practical applications of electricity were not discovered until late in the nineteenth century, when Thomas Edison, Nikola Tesla, George Westinghouse, and other inventors converted this curiosity into the products that underpin much of the world's modern economy. The first of those products, the incandescent light bulb, created demand for electricity.[15]

To meet that demand, the world's first electric power station opened in September 1881 in Godalming, a town in Surrey, England. The electricity was used to power thirty-four lights for the benefit of the public, but technical challenges and high costs led the town to abandon the system three years later.[16] The first commercial power plant in the world, New York City's Pearl Street Station, was built by Thomas Edison in 1882. The station used heat to run high-pressure steam engines which drove turbines to produce electricity. Edison burned coal to create heat in the steam engines because coal was inexpensive and readily available. By the 1880s, coal was already in widespread use to power steam engines used by factories, so it was logical to use coal in similar steam engines used to produce electricity.

The Pearl Street Station initially provided eighty-two customers with power for four hundred incandescent lights.[17] Customer demand grew rapidly, and the station powered over ten thousand lights a mere two years later. To Edison's chagrin, the Pearl Street Station was not a commercial success and burned down after only eight years in operation. But it demonstrated that there was customer demand for electricity, and Edison's design for coal-fired plants was replicated globally.

By 1900, the energy transition from wood to coal was complete among industrialized countries. In the United States and Europe, the great majority of energy used to heat homes, generate electricity, and drive industrial machines was derived from coal. Edison's original design for generating electricity from coal was so successful that it remained nearly unchanged for over one hundred years.

Beyond Coal

Until the middle of the twentieth century coal was the world's dominant source of energy for the generation of electricity.[18] But as with forests centuries before, coal mines in Britain became depleted, and as a result of increasing scarcity, prices rose. Along with higher costs, the effects of pollution from burning coal encouraged the search for new sources of power. Air quality in American and European cities in the 1950s and 1960s was shockingly bad. A "Great Smog" resulting from air pollution enveloped London for five days in 1952, killing up to twelve thousand residents and spurring the passage of environmental legislation.[19] Governments and industry began to search for an alternative to coal to generate electricity. Finding a superior source of energy required an understanding of the economics of electricity generation.

THE ECONOMICS OF ELECTRICITY

Electricity is a commodity, which means that consumers cannot distinguish among different sources of electrical power generation. The electrons generated from coal-fired plants are identical to those created by natural gas–fired or nuclear plants. Thus the primary differentiating factor among sources of electrical power generation is cost—the cheapest form of power production becomes the preferred source.

Comparing the cost of different sources of electrical power generation poses a challenge. Coal-fired power plants are relatively inexpensive to build, but an enormous amount of coal is required to generate electricity relative to other fossil fuels. Natural gas–fired power plants are significantly more expensive to build than coal, but they are less costly to run. Nuclear plants are very expensive to build, but very little uranium is required to power them relative to the inputs of coal and gas in facilities powered by those energy sources.

There are other costs to consider. Burning coal emits many pollutants, including asthma-causing soot and poisonous mercury. Natural gas also emits pollutants, but they are significantly fewer than those produced by coal. Nuclear power does not emit any pollutants, but the cost and risks associated with disposing of spent uranium, which is radioactive and highly toxic, can be colossal. How is one to compare these three very different sources of electrical power?

The solution is a formula, the levelized cost of electricity (LCOE), which is a standard metric used to compare the costs of producing electricity from different sources of generation. The LCOE provides an "apples-to-apples" comparison of cost. The LCOE for a power plant equals the cost of building and operating the plant divided by the electrical output forecast over the life of the plant, discounted at the cost of capital[20] required to invest in the plant.

The LCOE is calculated over the projected lifetime of the power plant, typically twenty to forty years, and is expressed in dollars per megawatt hour (MWh) of electricity produced (or in dollars per kilowatt hour [kWh], which is simply dollars per MWh divided by 1,000). The formula for calculating LCOE, and an example, can be found in appendix A.

LCOE allows for cost comparisons among very different sources of power using a single metric. This had important implications for the second energy transition, which was marked first by an attempted transition from coal to nuclear and then by a transition to natural gas.

THE ECONOMICS OF NUCLEAR POWER: A STALLED ENERGY TRANSITION

The concept of generating electricity from nuclear power was alluring as a replacement for coal-derived energy. Uranium is plentiful and cheap to mine and process, and the amount of energy created from splitting atoms is nearly unlimited. As President Dwight D. Eisenhower remarked in his "Atoms for Peace" speech to the United Nations General Assembly in December 1953, "Peaceful power from atomic energy is no dream of the future. The capability, already proved, is here today."[21]

Just four years after President Eisenhower's speech, the first large-scale nuclear reactor in the world began operations outside of Pittsburgh, Pennsylvania. Development of nuclear energy in the United States and in Europe boomed in the 1960s. In just one year—1973—U.S. utilities ordered forty-one nuclear power plants to be built,[22] and electricity generated by nuclear reactors in the United States grew from a near standing start to 20 percent of the total output by 1988.[23] The second energy transition was underway, this time from coal-fired generation to nuclear. Nuclear power's share of total energy generation grew quickly early on, and analysts predicted that it would dominate energy markets. Even U.S. naval ships were commissioned to run on nuclear energy.[24]

But the cost of building and operating nuclear plants began to climb when it became clear that radioactive waste disposal, safety features, and plant decommissioning needed to be included in the LCOE for nuclear energy. These additional costs were brought to light on March 28, 1979, when the nuclear power plant at Three Mile Island, in Pennsylvania, accidentally discharged large amounts of reactor coolant, leading to a partial meltdown and the release of radioactive gases. The cost of cleanup was $1 billion.[25]

Despite the momentum for nuclear to become the fuel of the future, the accident at Three Mile Island was a turning point in the growth of the U.S. nuclear power industry. Thereafter, regulators required additional safety measures, local opposition to nuclear plants became more strident, delaying project development, and construction costs ballooned. All of these dynamics contributed to a significant increase in the LCOE of nuclear power plants.

After the Three Mile Island incident, existing nuclear plants continued to operate, supplying 20 percent of the electricity generated in the United States in 2017, but the construction of new nuclear power plants never recovered.[26]

According to one commonly held belief, the demise of the nuclear industry is due to safety concerns and regulations. In fact, the failure of nuclear to replace coal is the result of simple economics. Nuclear power plants are an expensive way to generate electricity, as reflected in a high LCOE. The investment bank Lazard estimates the LCOE of nuclear power to be close to twice the unsubsidized LCOE of coal, natural gas, or renewable wind and solar power.[27] In 2017 alone, thirty-four of the sixty-one nuclear plants in the United States experienced a collective $2.9 billion in losses.[28] M. V. Ramana, a physicist at Princeton University, summed up the predicament of nuclear energy in an article in *Nature* in 2016: "The overwhelming factor shaping the future of nuclear power is its lack of economic competitiveness. Nuclear plants cost a lot to build and operate. This limits the rate of new reactor construction and causes utility companies to shut down reactors."[29]

The energy transition from coal to nuclear began with much promise but ultimately stalled due to uncompetitive costs. Nuclear power now plays a diminishing role in the generation of electricity globally, and there is little hope for a recovery.[30]

THE ECONOMICS OF NATURAL GAS: TRANSITION SUCCESS

An alternative source of energy, natural gas, was well known by Edison and other early inventors, but it was rarely used as a source of power because it could only be transported by pipeline.

Attempts were made in ancient China to transport natural gas using pipelines constructed of bamboo.[31] Engineers subsequently tried using wood and other nonmetal materials, none of which were capable of efficiently transporting economic volumes of gas. In the late 1920s, improvements in welding techniques and use of metals to create pipes led to pipeline construction that was more economically viable than in the past. In the United States, natural gas pipelines were constructed in the 1930s and 1940s, allowing for the transmission of natural gas for lighting, heating, and cooking, and for industrial applications. By this time "the U.S. pipeline network, laid end-to-end, would stretch to the moon and back twice."[32] But natural gas was not widely used in electricity generation until the invention of efficient gas turbines.

In the 1990s, development of new and better gas turbine technology, combined-cycle gas turbines, increased power generation from plants fired by natural gas by capturing both heat at generation and waste heat to improve efficiency and electrical output (hence the term "combined cycle"). In the United States, growth in demand for natural gas was in part the result of deregulation of natural gas prices in 1990, part of a broader deregulation trend that started in the late 1980s. In the following decade, Europe's interest in lower-emissions energy production, combined with Asia's thirst for quickly-installable energy plants to meet the needs of its rapidly growing population, meant that gas turbine equipment was in high demand.[33]

By 2000, natural gas had surpassed coal as the preferred source of new electricity generation. But increasing demand for natural gas drove up prices, which threatened to derail the transition from coal. The technological innovation by which natural gas was extracted using a technique called hydraulic fracturing, more commonly referred to as fracking, solved the problem of growing demand for natural gas.

Fracking uses high levels of pressure to force open fissures in shale rock, releasing natural gas. Fracking was first discovered during the Civil War, but the modern fracking boom began with George Mitchell, a Texan oilman.[34] Mitchell was famously stubborn, saying "I never considered giving up" despite years of failure.[35] Experimenting over several decades, he perfected the fracking process, combining several technologies, notably the use of more effective fluids and horizontal drilling, to dramatically lower the cost of extracting natural gas. Mitchell's eventual success paved the way for rapid growth in the U.S. shale gas industry, in which production volumes boomed while prices remained low.

Buoyed by abundant natural gas at low prices, construction of natural gas–fired power plants in the United States totaled 228 GW from 2000 to 2015, compared to only 20 GW of new coal-fired capacity.[36] The U.S. Energy Information Administration concluded this was due to the "cost competitiveness of natural gas relative to coal."[37] The lower LCOE of electricity generated using advanced combined-cycle natural gas technologies encouraged utilities to transition away from coal. By 2016, natural gas had surpassed coal in the generation of electricity in the United States.[38]

THE ECONOMICS OF ELECTRICITY GENERATION AT THE BEGINNING OF THE TWENTY-FIRST CENTURY

Energy was first derived from the burning of wood, a cheap and abundant resource for thousands of years. The development of industrial machinery, and the demand for electricity to power those machines and to light homes and buildings,

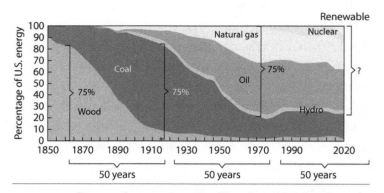

FIGURE 2.2 Sources of energy, 1850–2020. Note that oil is primarily used for transportation; coal, natural gas, nuclear, and renewables are primarily used to generate electricity. (Adapted from U.S. Energy Information Administration [EIA], *Annual Energy Review 2008*; and EIA, *Annual Energy Outlook 2009, with Recovery Act Update*)

forced the first energy transition to coal—a relatively energy-dense source of power that was abundant and cheap. Coal underpinned the Industrial Revolution. However, by the 1950s the limitations of coal, including the damage caused by smog and other pollution, gave way to the advantages of generating electricity from nuclear power and natural gas. But nuclear power turned out to be an expensive source of power generation, and the transition to nuclear stalled. Natural gas, on the other hand, had a low LCOE and thus became the preferred source of new power generation. By the beginning of the twenty-first century there had been a partial transition in electricity generation, first from coal to nuclear power, and then to natural gas. On the horizon lay another energy transition, from fossil fuels and nuclear to the most common source of energy on the planet—the sun.

ENERGY TRANSITIONS:
LESSONS FROM THE PAST

The transitions from wood to coal and from coal to nuclear and natural gas highlight four enduring lessons about energy transitions. The first and most important lesson is that basic economic principles, primarily cost, are the main drivers of energy transitions. Cost is key because one form of energy can often be substituted for another, especially in the generation of electricity. The second lesson partially contradicts the first—despite lower cost, energy transitions are slow and often delayed due to regulations, the sway of incumbency, or uncertainty about new technology. The cheaper source of energy is not readily accepted by those in power, such as the nobility in Victorian England, slowing the transition. The third lesson is that innovation, especially in the form of technology, can trigger or accelerate energy transitions, just as Watt's steam engine hastened the transition to coal. The fourth and perhaps most interesting lesson is that energy transitions have unforeseen but dramatic material consequences. The transition from wood to coal underpinned the Industrial Revolution and the dramatic increase in economic growth and human welfare that followed. But it also began the process of anthropogenic (man-made) climate change. These consequences—positive and negative—will be explored at the end of this book.

FIGURE 3.1 The Three Gorges Dam. (Source: Wikimedia Commons)

3

THE RISE OF RENEWABLES

SCIENTISTS have long aspired to create power from renewable sources of energy. By 2015, that aim had been realized in the commercial realm, with annual investments in renewable energy more than double the investments in power generation by fossil fuels.[1] Remarkably, over five hundred thousand solar panels are installed globally every day; in China, two new wind turbines are installed every hour.[2] And the future looks bright. Analysts predict that renewable energy plants will make up nearly three-quarters of the $10.2 trillion invested in new power generation between 2017 and 2040.[3] What created the transition to renewable energy? The answer is cost competitiveness. Renewable energy, in the form of wind and solar power, has become cost competitive with alternative forms of electrical power generation. How that occurred is the subject of the next three chapters.

RENEWABLE ENERGY SOURCES

Energy from a source that is not depleted when used is considered renewable energy. This includes energy generated from the wind and the sun; from rivers, tides and waves; and from the earth's internal heat. It also includes the use of biomass and biofuels created from plant matter, as these can be regrown. These energy sources are all naturally occurring and theoretically inexhaustible.

ENERGY FROM WATER

Energy generated from hydropower is historically the most important source of renewable energy. During the industrial revolution textile mills were often powered by hydro, and the first hydroelectric power plant was built in 1882. The first mega hydroelectric project, the Hoover Dam on the Colorado River, was completed in 1936, with the capacity to generate 1,345 MW of electricity. At the time, it was the world's largest power generation project. Hydro construction reached a new milestone in 2012 with completion of the Three Gorges Dam in China. The scale of Three Gorges is immense, capable of generating 22,500 MW of power, sixteen times the Hoover Dam. To put that into perspective, construction of the Three Gorges Dam required the permanent relocation of 1.2 million people.[4]

Three Gorges was followed by development of several other mega hydro projects in China and Brazil as well as of hundreds of smaller projects elsewhere. However, the potential of hydro in the future is limited, as most of the major rivers in the world have now been dammed for hydroelectric power. In 159 countries, hydropower is considered a "fully mature technology."[5] In the United States, hydropower peaked as a percentage of electricity generation at 20 percent in 1974, and it has not grown since then.[6] Globally, hydropower's share of electricity production also peaked at 20 percent in 1974 and has declined to 16 percent, and this figure includes electricity generated from massive hydropower projects like the Three Gorges Dam.[7] Hydropower will remain an important source of renewable energy, but this technology has fewer locations to scale while remaining cost competitive with other forms of power generation.

Generating power from water also includes tidal and wave power. Unlike hydropower, tidal and wave power remain experimental, with limited locations to scale and high costs. Theoretically, any country with a coastline could deploy resources to capture energy generated from large bodies of water, but at present just two nations, France and South Korea, are responsible for 90 percent of the world's tidal and wave energy generation. Waves have the benefit of being more energy dense than wind, which helps explain why, according to researchers at Oregon State University, just 0.2 percent of the ocean's untapped wave energy could power the planet.[8] As this chapter explores, however, *potential* is only part of the equation. The cost of generating electricity from tidal or wave power remains much higher than most other forms of power generation, severely limiting its actual use.

ENERGY FROM PLANTS

The burning of biomass, primarily wood, has supplied energy since humankind's earliest days. Even today, wood is the primary form of fuel for cooking in many developing countries. Wood pellets are also burned for heating in some modern homes, and a few advanced countries use biomass for a significant share of their power. In Sweden, 22 percent of the country's energy supply comes from wood.[9] But burning wood is, in most countries, relatively expensive and inefficient. Wood is not nearly as energy dense as fossil fuels, it is costly to transport, and, while forests are renewable, they are easily depleted and take decades to recover.

As an alternative to burning biomass, energy from plants can also be generated with biofuels. The most common forms of biofuel are ethanol and biodiesel. Ethanol is an alcohol produced by fermenting biomass, a process similar to the brewing of beer. Biodiesel is created through a chemical process that separates vegetable or fat oil from biomass. Ethanol can be blended with or replace gasoline, and biodiesel can replace diesel fuel. Both ethanol and biodiesel offer simple, proven techniques for the replacement of fossil fuels with renewable energy in the transportation sector.

Turning plant biomass into biofuel is theoretically possible anywhere, since biomass is technically defined as any living plant or animal.[10] In reality, few countries have scaled biofuel production, as the economics have turned out to be quite unfavorable. Liquid fuel in the form of ethanol from corn and sugar cane can be used for transportation and replaces some demand for petroleum products. But the inputs to biofuel must be grown, and this requires land, water, and other resources. For example, since 2011 more of the corn grown in the United States—the largest producer of biofuel globally—is used for biofuel than for food.[11]

Furthermore, turning living matter into fuel is inefficient. The World Resources Institute (WRI) calculates that "providing just 10 percent of the world's liquid transportation fuel in the year 2050 would require nearly 30 percent of all the energy in a year's worth of crops the world produces today." The WRI analysis goes on to suggest that "fast-growing sugarcane on highly fertile land in the tropics converts only around 0.5 percent of solar radiation into sugar, and only around 0.2 percent ultimately into ethanol." Right behind the United States as a producer of biofuel is Brazil, where sugarcane waste is converted into ethanol. Yet even in Brazil, which meets the "best conditions" outlined by the WRI, production of biofuel is simply not an efficient method for capturing energy from the sun and converting it into useable energy on earth. The WRI study found that solar power systems can produce more than a hundred times more useable energy per hectare than biofuels.[12]

Biofuels can also be produced from cellulosic material, such as wood chips, and from algae and other aquatic plant material. Referred to as second generation biofuels, these renewable feedstocks are used to produce energy more efficiently than corn or sugarcane. Unfortunately, despite large investments of capital and engineering expertise, commercial production of these biofuels remains a distant proposition. Absent a dramatic breakthrough in technology, second-generation biofuels are likely to remain uncompetitive with petroleum and other fossil fuels, especially for energy use on a global scale.

ENERGY FROM THE EARTH

Another renewable source of power, geothermal energy, takes advantage of energy already contained within the earth. Geothermal power is harnessed from the heat within the earth, which was created during the planet's formation. Like hydropower, geothermal can be cost competitive with fossil fuels. Unfortunately, in another parallel to hydropower, geothermal has limited future potential, as the most easily accessible sources of geothermal power have already been built.

Geothermal power accounts for a mere 0.3 percent of global power generation, since it can only be harnessed under specific geologic conditions, although in some countries geothermal accounts for much more (in Kenya, which boasts the world's largest geothermal plant, it accounts for 44 percent of energy generation).[13] Geothermal power has one of the lowest LCOE ranges of all energy sources, although it is capital intensive to conduct seismic and other exploratory testing and to construct plants. Geothermal power is reliable and consistent for generating electricity. Unfortunately there are a limited number of places in the world where geologic conditions make it both accessible and cost competitive.[14]

RENEWABLES FOR AN ENERGY TRANSITION

Energy generated from hydropower, tidal, wave, biomass, biofuels, and geothermal contributes to the production of power globally, and each of these energy sources is renewable. Yet these technologies suffer from high cost, limited growth opportunities, or both, which severely limits the potential for any of these renewable energy sources to replace fossil fuels. These renewable sources provide useful energy, but given the currently available technologies, they are ill suited to lead a global energy transition.

This leaves two remaining sources of renewable energy for consideration: wind and solar. Both power sources are unlimited, globally abundant, and perpetually renewable. And, as will be explained in the next two chapters, both wind and solar power are low cost. Despite the many sources of renewable energy, the remainder of this book focuses on only wind and solar. Those are the only sources of renewable energy that are both cost competitive with fossil fuels *and* have the potential to scale globally.

FIGURE 4.1 Smoky Hills Wind Farm, Kansas, United States.
(Source: Wikimedia Commons)

RENEWABLE WIND ENERGY

HUMANS have harnessed energy from the wind ever since
the ancient Persians developed vertical-axis windmills
in 200 BC, which produced mechanical power to grind
grain. In the United States, early settlers built windmills throughout the countryside to pump water for irrigation and, eventually, to power homes and businesses. But the use of windmills
declined with the advent of the modern electrical grid, which
provided farmers with reliable electricity at low cost.

The global energy crisis of 1973 prompted many governments
to renew interest and research in wind turbines. Denmark was
an early leader in the sector due to the research of Poul La Cour,
a scientist, inventor, and early pioneer of wind power in the late
1800s.[1] Danish innovators picked up where La Cour left off, and
by 1979 Denmark had installed its first modern wind turbine,
setting the stage for its global leadership in turbine development
and installation.[2]

An early windmill designed by one of La Cour's students had
blades constructed of wooden frames and covered with aluminum alloy sheets.[3] Modern wind turbines are constructed of steel,
aluminum, copper, and composite materials, including plastic
reinforced by glass and carbon filament, wood epoxy, prestressed

concrete, and magnetic materials. This approach brings together a complex design and advanced materials with a simple objective: to build bigger.[4]

BIGGER IS BETTER

Wind turbines create power by capturing the energy in the wind. Simple physics explains why turbine height and blade size are the primary factors for determining electricity production from wind power. The amount of energy captured is a function of the speed of the wind and the area that is swept by the blades of the turbine. Wind speed is especially important, as the amount of energy in the wind is proportional to the wind speed cubed. For example, winds at 20 mph have *eight* times the energy of winds at 10 mph. In other words, double the wind speed and the energy output increases eight times.[5] Importantly, wind speeds are stronger at greater heights, meaning that a taller wind turbine is exposed to more energy available for capture than a shorter one.

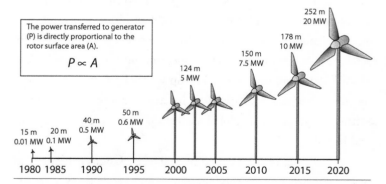

FIGURE 4.2 Relationship between wind turbine size and power. (Adapted from the European Wind Energy Association, *Wind Energy—the Facts: A Guide to the Technology, Economics, and Future of Wind Power*, 2009)

Just as the energy in wind is proportional to speed, the area swept by the blades is proportional to the size of the wind turbine's blades. Area equals πr^2, where r is the length of the wind turbine's blade. A blade that is 12 meters in length will sweep an area that is *four* times as large as a blade that is 6 meters in length. Consequently, increasing the height of a wind turbine enough to double the blade length will quadruple the swept area of the blades. And a taller wind turbine will expose those blades to higher wind speeds. The combination of longer blades and taller turbines will result in a dramatic increase in energy captured.

It is therefore not surprising that the history of modern wind turbines is one of increasingly bigger structures. The turbines developed in the 1980s had 15-meter rotor diameters and generated 50 kW of power, enough to meet the electricity demands of about ten homes. By 2005, innovations in materials and technologies allowed for the construction of wind turbines with rotor diameters of 124 meters, generating 5,000 kW (or 5 MW) of power, enough for a single wind turbine to power a thousand homes. The innovations and advances in technology that created bigger turbines resulted in a hundredfold increase in power generation.[6]

THE ECONOMICS OF WIND POWER

The economics of building and operating a wind farm are determined primarily by the capital costs of the wind turbines and by their installation and interconnection to the grid to transmit the electricity generated by the facility. The largest capital cost is the wind turbine itself, which is composed of a tower, a nacelle which houses the equipment at the top of the tower, and the blades. The tower is relatively simple in design and is usually constructed of steel. The components of the nacelle are complex, designed to both optimize electricity generation and control the turbine in a wide range of wind speeds. Modern wind turbines use advanced gearing systems to maximize power generation and an active control system that keeps the blades pointing in the direction of the wind. The braking system prevents damage to the turbine when wind speeds become excessive—during a storm, for example. The blades are, of course, the critical component for capturing energy from the wind. Modern wind turbine blades are composed of advanced composite materials designed to extract as much energy as possible yet lightweight and durable in a wide range of wind speeds.

Installation of a wind turbine is straightforward, although transport of the turbine from the factory to the wind farm site can be challenging due to the immense size of the various components. A large wind turbine is significantly bigger than a jumbo jet, so special-purpose ships and trucks are required to transport the turbines. Once it arrives on site, the wind turbine is placed on a steel-reinforced cement foundation and connected to the electrical grid; this installation is called an interconnection. The cost of the interconnection is mostly dependent on the distance between the wind farm and the nearest transmission lines.

A wind turbine has a rated power output, expressed in megawatts, which represents the maximum amount of power that can be generated by the turbine at any given point in time. On modern wind farms, turbines range from one to eight megawatts in size. The electricity generated by a wind turbine is the rated power output in megawatts multiplied by the number of hours of operation. Of course, wind turbines only generate electricity when the wind is strong enough to turn the blades, typically this means a wind speed greater than 7 mph.[7]

The most important calculation for determining the economics of a wind farm is the capacity factor, defined as the ratio of actual power output to rated power capacity if the turbines were always operating. Onshore wind farms experience average capacity factors of 35 percent, meaning that the electricity generated is 35 percent of the rated power output.[8] In other words, the wind is strong enough to turn the blades an average of 35 percent of the wind farm's rated output throughout the year. For example, a 5-MW wind turbine operating with a 35 percent capacity factor will generate 15,330 MWh of electricity per year.[9] Note that offshore wind projects experience higher capacity factors, as wind speeds are typically higher and steadier over the water.

The operating costs of a wind farm are primarily the costs of leasing the land on which the turbines are located (described in more detail below), operating the project, and maintenance. Operating a wind farm requires little oversight or expense, as wind turbines are highly automated. Maintenance, however, can be costly. Wind turbines are designed to generate electricity for more than twenty years, but they are mechanical devices, which means that components eventually wear out. The first modern wind turbines experienced frequent failure in the gear boxes, contributing to high repair costs. Subsequent improvements in engineering design led to gearless turbines which are more reliable and therefore less costly to maintain than earlier models.[10]

In the 1980s and 1990s the economics of wind farms were quite poor. Wind turbines were small and therefore had low power ratings. The high capital costs of wind turbines meant that wind farms were expensive to build, and the low power ratings meant they generated little electricity. In comparison to electricity generated from coal, nuclear, and natural gas, early wind farms had a relatively high LCOE, making them uncompetitive with traditional forms of energy.

Government subsidies, especially in Europe, provided incentives for wind developers to build wind farms, creating demand for wind turbines from manufacturing companies. This created a tipping point, as government incentives compensated for the relatively poor economics of wind, allowing manufacturers and project developers to build wind projects. This, in turn, created expertise in the sector. As expertise improved, costs declined, and more wind projects were built. But wind energy companies knew that government subsidies would not last forever. The poor economics of wind energy forced the industry to focus on lowering the LCOE of wind to compete with traditional forms of electricity generation.

DECREASING THE LCOE
OF WIND POWER

The physics behind generating energy from the wind led to the development of increasingly bigger and more efficient wind turbines. But when generating electricity, bigger is only better if it is also cheaper. The modern wind industry, therefore, focused relentlessly on reducing the LCOE of wind power to improve competitiveness with other forms of electricity generation. It worked. For wind power, the LCOE has declined from an average of $500 per MWh in the 1980s to $45 per MWh in 2017.[11] In other words, the cost of electricity generated by wind turbines declined by more than 90 percent, becoming competitive with electricity generated by coal, nuclear, and natural gas. How was this possible?

THE LEARNING CURVE

The process by which costs are reduced as production volumes increase is referred to as the learning curve, an economic concept that has significant implications for the growth of many forms of renewable energy, including wind power. The learning curve measures the ability of an industry to improve the performance or reduce the cost of products as the volume of production increases. The theory underlying the learning curve is that consistent improvement in performance or cost is possible through increased experience. Moore's law is the best-known application of the learning curve. Gordon Moore, the cofounder of Intel, forecast that the capacity of computer chips would double every two years, implying a learning curve of 40 percent, a forecast that has held roughly true for fifty years.[12]

It should be noted that the learning curve as it applies to advances in renewables is not time bound like Moore's law. It is simply a way to demonstrate the rate of technological improvements relative to production volume, regardless of how long those improvements take. In renewable energy, the learning curve specifically measures the percentage change in LCOE each time the total installed capacity is doubled.

Manufacturers of wind turbines and developers of wind farms are constantly improving production techniques, applying new technologies, and taking advantage of economies of scale. And each time such a development occurs, wind energy companies learn how to reduce costs and increase output, thereby reducing the LCOE of wind power. The installed capacity of wind turbines has been doubling every four to five years, and as production has increased costs have decreased.

For example, the total installed capacity of wind turbines approximately doubled from 238 GW to 487 GW[13] between 2011 and 2016.[14] During this same period, the average LCOE of wind declined from $71 per MWh to $47 per MWh.[15] This means that the learning curve of wind during that period was 34 percent. In other words, doubling wind power installations resulted in a decrease of 34 percent in the LCOE of electricity generated by wind power. Academic studies of the wind power industry in Europe and the United States confirm that the learning curve in Europe is significant, if slightly lower than in the United States, with estimates ranging from 11 percent to 19 percent.[16]

The learning curve creates a virtuous cycle for emerging industries such as wind and, as the next chapter will explore, solar. The virtuous cycle works as follows: each time sales increase, production volumes rise, manufacturers learn from experience, and costs decrease. The cost decrease results in a further increase in sales and production, and in even lower costs, relentlessly driving down the LCOE of wind power.

Consistent declines in the LCOE of wind power have driven rapid growth in wind projects. Fortunately, traditional farms provide ideal locations for modern wind farms.

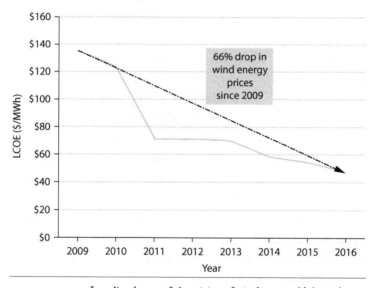

FIGURE 4.3 Levelized cost of electricity of wind power. (Adapted from Mike O'Boyle, "Wind and Solar Are Our Cheapest Electricity Sources — Now What Do We Do?," America's Power Plan, December 21, 2016. Data from Lazard's *Levelized Cost of Energy Analysis 10.0*)

FARMING THE WIND

A prime location for modern wind turbines is in fields used for agriculture. Placing wind turbines on traditional farms generates significant incremental income for farmers who lease the land to the wind project owners. Each turbine requires approximately 50 to 100 acres of land, as the turbines must be spaced far apart so that wind turbulence from one turbine does not affect the performance of others.[17] Despite their large size, the bases of wind turbines are small. The largest wind turbines require less than an acre of land for the foundation. This means farmers can continue to use land they have leased to wind farms for grazing livestock or growing crops. In the United States, farmers typically earn $7,000 to $10,000 per turbine annually simply by leasing the land to the wind turbine owner,[18] dramatically more than what they earn from farming the equivalent land area.[19] Earnings from wind turbines provide a buffer against fluctuating agricultural commodity prices, as leases are fixed for twenty years or more. Optimal conditions for wind farms are often found in areas most in need of the additional income: 70 percent of U.S. wind farms are in low-income, rural communities.[20] Additionally, the increase in land values from wind projects results in more tax revenue for state and local municipalities. According to a farmer in Iowa, "This is our financial future."[21] But wind energy is not without its challenges, and foremost among these is its intermittency.

CHALLENGES IN THE TRANSITION TO WIND POWER

To state the obvious, wind turbines only generate electricity when there is sufficient wind to turn the turbine blades. On most wind turbines, this requires a minimum wind speed of 7 mph.[22] At low wind speeds, turbines do not generate any power, and even moderate winds do not optimize electricity generation. Overall, U.S. wind farms experience capacity factors averaging 35 percent.[23]

Even at the windiest sites, such as West Texas or the North Sea, capacity factors are below 50 percent, meaning that wind turbines generate electricity for fewer than twelve hours per day. Furthermore, generation is not consistent: it can change hourly or even minute by minute. Wind power is therefore considered intermittent—it can only be produced when the wind is blowing. This is in contrast to electricity generation from fossil fuels, nuclear, and hydro, which are dispatchable, meaning electricity can be generated whenever required. The intermittency of wind is only a minor challenge when wind power accounts for a relatively small percentage of all electricity on the electrical grid, but it becomes a much more significant challenge as production increases. Intermittency, and solutions to the challenges it poses, are explored in greater depth in chapter 10.

Wind farms face additional hurdles, mostly related to siting (the geographic placement of wind turbines) and transmission of the electricity generated by turbines to the electrical grid. Wind turbines are popular with consumers seeking lower electricity costs and cleaner energy, and among farmers and other landowners receiving annual payments for leasing small plots of land. Their neighbors are not always as keen. "Not-in-my-back-yardism" ("NIMBYism") is a chronic challenge to the development of wind farms.

The classic example of NIMBYism is the Cape Wind project, which if developed will be visible on the horizon to residents of Cape Cod, Massachusetts. The project was initially proposed in 2001, with the support of 55 percent of voters on Cape Cod[24] as well as many local newspaper boards, elected officials, and civic associations, but it was opposed by a small coalition of wealthy residents concerned that the wind farm would affect their views and potentially depress their property values. In what was perhaps the epitome of NIMBY, former news anchor and Martha's Vineyard resident Walter Cronkite, an early opponent of the wind farm, said of the project: "There must be other locations to make it possible."[25] Development of the Cape Wind project has been delayed by a small number of vocal opponents for more than seventeen years and may never be built, demonstrating the challenge of NIMBYism to wind energy.

An additional siting challenge for wind farms is the need to transmit the electricity generated by the wind turbines to the electrical grid, where it is used by homeowners and businesses. The optimal locations for wind farms are those where it is windiest; naturally, many of those places are far from where people live. In the United States, the windiest states are Nebraska, Kansas, and the Dakotas, none of which has significant population centers within those states or nearby.[26] In Europe, the windiest locations are offshore, in the North Sea. As such, wind farms often require new transmission lines to move electricity from its source to where it will be used. Transmission lines are constructed at considerable cost. Moreover, electrical losses increase as the distance from generation to consumption increases, further increasing the price of the power.

Despite these challenges, wind power has grown rapidly, and the sight of wind turbines is increasingly common. Development of wind farms is sometimes delayed by neighbors, and the cost of electrical transmission can prohibit construction in remote areas, but the increasingly attractive economics of wind power inevitably overcome these hurdles.

WIND POWER IN THE
TWENTY-FIRST CENTURY

Wind power, the generation of electricity from wind turbines, is an increasingly competitive source of energy. Relentless advances in technology combined with targeted government incentives have created a virtuous cycle of lower costs, growing demand, and a steadily declining LCOE for electricity. The learning curve of manufacturers and developers continues to drive the LCOE of wind lower, ruthlessly improving the cost competitiveness of wind power against other forms of electricity generation. The results have been impressive. By 2015, wind farms were producing 5 percent of electricity in the United States and 10 percent of electricity in Europe; those figures had been near zero a decade earlier.[27]

FIGURE 5.1 Albert Einstein. (Source: Wikimedia Commons)

5

RENEWABLE SOLAR ENERGY

O
UR sun is the source of all energy on earth, warming the planet and making life possible. Photosynthesis converts sunlight to energy in plants, which can later be recovered indirectly through burning wood and other plant matter or by burning the organic matter which has been converted through geologic processes into fossil fuels. Sunlight provides as much energy in ninety minutes as is consumed by every person on the planet in a year.[1] Unsurprisingly, scientists have long dreamed of capturing that energy directly.

In 1905, Albert Einstein published the first paper describing the photoelectric effect, positing that light contains packets of energy, which he called photons. When metals are exposed to these photons, they emit electrons, which can then be harnessed for electricity. This theory was radical at the time, propelling Einstein out of obscurity and laying the foundations for quantum theory. For this paper, Einstein later received the Nobel Prize in physics. But it took another half century for scientists working at Bell Labs to develop the first functional photovoltaic solar cell, praised by the *New York Times* as "the beginning of a new era, leading eventually to the realization of harnessing the almost limitless energy of the sun for the uses of civilization."[2]

The first practical application of solar energy was on the Vanguard 1 satellite launched by the United States in 1958. It was an auspicious start, as the solar panel operated continuously for seven years, substantially longer than the satellite's conventional batteries, which lasted only twenty days.[3] NASA went on to use solar panels on both satellites and spacecraft. While the use of solar panels flourished on space missions, they were considered far too expensive for most applications back on earth.

Slowly but surely, solar energy spread, from space-based applications to locations in the arctic, on drilling rigs, and remote islands. The energy crisis of the late 1970s encouraged companies to develop better performing solar products at lower cost. But growth in the solar sector was slow. By the end of the twentieth century, total power generation from installed photovoltaic systems globally had reached only 1 GW, equivalent to the electricity that could be produced by just one coal- or natural gas–fired power plant. Nearly a century after Einstein published his paper on the photoelectric effect, generating electricity from the sun was confined to a few niche applications; the simple reason for this was cost.

THE ECONOMICS OF SOLAR POWER

The raw materials required to produce electricity from light are abundant and inexpensive to source. Unfortunately, the manufacturing process is complex and costly. Photovoltaic (PV) systems are composed of solar panels or modules, each containing many solar cells. Solar energy, the generation of electricity from light, occurs when the solar cells convert light into electricity using semiconducting materials. In most solar cells, the semiconducting material is silicon, one of the most abundant materials on Earth, and the same material used in computer chips. When light is absorbed by the silicon semiconductor, the energy in light photons moves electrons which then flow as electrical current through the solar cell along wire conductors. PV systems can also be built using semiconducting materials other than silicon (one example is cadmium telluride); 90 percent of PV panels, however, utilize silicon.

Not all the sunlight reaching the solar cell is converted to electricity. Conversion efficiency is the ratio between the useful output of an energy conversion device and the input. The maximum conversion efficiency of solar PV panels is 29 percent in theory, and in practice the maximum conversion efficiency achieved is 26.3 percent.[4] Most commercial PV panels have conversion efficiencies in the low 20-percent range. This has important implications for the cost of electricity produced by solar energy, as higher conversion efficiencies produce more electricity but are more expensive to manufacture.

The cost of manufacturing solar PV panels is measured in dollars per watt. For example, a panel than can produce 200 watts of electricity, with a manufacturing cost of $600, has a cost per watt of $3.[5] But cost per watt is not the most important economic measure for energy; as with wind, the levelized cost of electricity (LCOE) is what really matters. To calculate the LCOE of solar energy, it is necessary to determine capital costs to install the PV system, operating costs over the life of the system, expected output of electricity, and the cost of capital to finance the project.

Capital costs include the cost of panels and what are referred to as the balance of systems (BoS), which include mounting racks to hold the panels, an inverter to convert electricity generated from direct current (DC) to alternating current (AC), electric cables, and monitors. Capital costs may be reduced by government subsidies or incentives, discussed in more detail in the next chapter. BoS costs and installation labor account for 50 percent or more of solar system costs, while the PV panels make up the other 50 percent.[6]

Operating costs are very low, as PV systems contain no moving parts and last for thirty years or more. The primary operating cost is siting, which is the use of the rooftop or land on which the solar panels are located. PV systems mounted on homes or other buildings do not often incur siting costs, but larger solar projects installed in fields require significant lease payments, as the land cannot be used simultaneously for farming or other purposes.

In the early days of the solar industry, when PV panels were used primarily by NASA on spacecraft, the cost per watt was over $100. By 2000, that figure had declined to $5, an impressive reduction in cost.[7] Even this was not low enough for electricity from solar to have an LCOE competitive with other sources of electricity generation, except in niche applications. But each time the price of PV panels declined, the number of niche applications increased, and production of panels expanded. As manufacturing of PV panels increased, costs further declined, creating a virtuous cycle. By 2017, the cost of producing a solar PV panel had plummeted to $0.29 per watt, a 94 percent cost reduction in only seventeen years.[8] The plunge in panel prices resulted in a remarkable decline in LCOE for electricity generated using solar PV, making solar energy competitive with other sources of electricity generation, including fossil fuels. Once again, the learning curve was key to this transformation.

THE LEARNING CURVE APPLIED TO SOLAR POWER

Recall that the learning curve measures the ability of an industry to improve the performance or reduce the cost of products as the volume of production increases. The learning curve is calculated

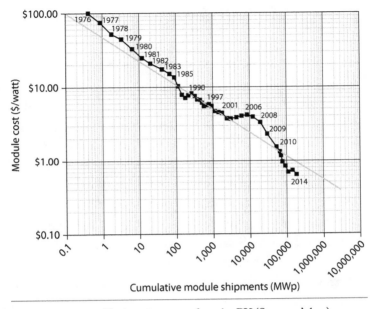

FIGURE 5.2 The learning curve for solar PV (Swanson's law).
(Source: Wikimedia Commons)

as the percentage drop in price for each doubling of cumulative production. Richard Swanson, the founder of solar panel manufacturer SunPower Corporation, observed that the cost of producing solar PV panels declines approximately 20 percent for each doubling in cumulative production.[9] This observation became known as Swanson's law.

Academic research has confirmed that Swanson's law is reasonably accurate, as learning curves for solar PV systems had a mean of 23 percent across multiple studies.[10] Those studies found that the learning curve works well for solar PV for several reasons, including economies of scale, improvements in conversion efficiency, and advances in manufacturing processes.

LCOE OF SOLAR POWER

Swanson's law has important implications for the LCOE of solar energy. As the solar industry grew, the price of solar panels declined, and since solar panels make up approximately half of the capital costs of a solar project, the overall LCOE declined as well. But the declining cost of PV panels is only part of the story; understanding the LCOE of solar energy first requires an understanding of the different types of solar energy projects.

Electricity from solar panels is commonly referred to as distributed generation, as solar panels can be placed in almost any location and at any scale—from a single panel up to thousands of connected panels. Solar panels can be used to power everything from a hand-held calculator to a house, an entire building, or even a city. This stands in contrast to the traditional form of electricity generation, originally developed by Thomas Edison and used in his Pearl Street Station, of centralized generation, in which a large power plant generates power that is transmitted to many users across the electric grid. Distributed power generation has much greater flexibility in design than centralized power generation, allowing for several different categories of solar projects.

Solar projects are typically classified in three different categories: residential, commercial, and utility. A fourth classification, off-grid, refers to solar projects that are not connected to the electrical grid; these are found primarily in developing countries.

Residential solar projects are mounted on an individual home, with the electricity generated from the solar panels used by the homeowner (note that homeowners can also install passive solar systems for heating water, a completely different technology). The average residential solar project is composed of twenty solar panels, with a capacity of 250 watts per panel, generating 5 kW of power when the sun is shining, enough to satisfy the electrical needs of the average-sized American or European house.[11]

Residential solar projects are referred to as "behind-the-meter," as the electricity from the solar panels is used where it is generated and does not pass through the local electric grid. If the solar panels generate more electricity than can be used in the home, the excess electricity then flows through the electrical meter, out of the house, to the electric grid. In most cases, home-owners with solar systems generating excess electricity receive a financial credit from the grid operator or utility for sending electricity to the grid. This is referred to as net metering. Residential-scale solar constitutes by far the largest number of solar projects globally, although each project is very small. Approximately one-quarter of all electricity generated from solar is from residential solar projects.[12]

Commercial projects are defined as projects supplying electricity to buildings or businesses. Like residential solar, most commercial scale projects are located on building rooftops, though occasionally the projects are located on land next to the building. Commercial solar projects average 500 kW in size, or approximately two thousand solar panels.[13] Again, like residential solar, commercial solar projects generate electricity that is used "behind-the-meter" in the building. The electricity generated in a commercial scale project is sometimes shared by multiple tenants or businesses in a single building. Like residential solar, approximately one-quarter of all electricity from solar is from commercial projects.[14]

Solar projects that are large in scale and generate electricity explicitly for distribution to the electrical grid are referred to as utility scale. The solar PV panels used in utility-scale projects are always mounted on the ground, usually in fields. To provide perspective, the smallest utility-scale solar PV projects are 1 MW in size, composed of approximately four thousand panels on seven acres of land. The largest utility-scale projects built to date are composed of several million solar panels on thousands of acres of land.

Utility-scale solar projects are normally composed of photovoltaic panels, but they can also be constructed using a very different technology called concentrated solar power (CSP). CSP solar plants utilize mirrors or other reflecting lenses to concentrate the sun's rays, where it is converted to heat and used to drive a steam-powered turbine to generate electricity. Several dozen large CSP projects have been constructed and are operational, primarily in Spain and the United States.[15] However, concentrated solar power has fallen out of favor because the cost of generating electricity from CSP is significantly higher than from PV projects, making it less competitive on an LCOE basis.[16]

The solar PV panels used by all three categories of solar projects—residential, commercial, and utility—are identical.

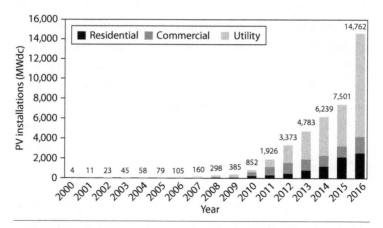

FIGURE 5.3 Annual U.S. solar PV installations. (Adapted from GTM Research and Solar Energy Industries Association, *Solar Market Insight Report: 2016 Year in Review*)

The only differences among the projects are the number of panels used and whether they are installed on a roof or on the ground.

In addition to the cost of panels and balance of systems, solar projects also incur soft costs. This includes the cost of sourcing and contracting with the site owner, permitting, indirect labor, overhead, and margin. Unsurprisingly, residential solar projects have high soft costs, as each home requires a separate set of contracts, permits, and so on, and they do not benefit from economies of scale. Figure 5.4 demonstrates that the total cost of developing solar projects varies dramatically across the different project categories, with the cost of residential projects nearly three times that of utility-scale projects on a per watt basis. Utility-scale solar projects enjoy economies of scale that are better than commercial projects and markedly better than residential projects.

The dramatic decline in solar PV panel prices has contributed to lower costs for the entire solar power sector. The unsubsidized LCOE of solar power from utility scale projects has declined

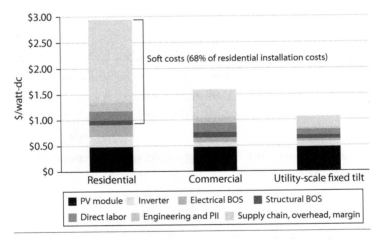

FIGURE 5.4 Solar PV prices. (Adapted from GTM Research, Solar Energy Industries Association, *Solar Market Insight Report 2016 Year in Review*)

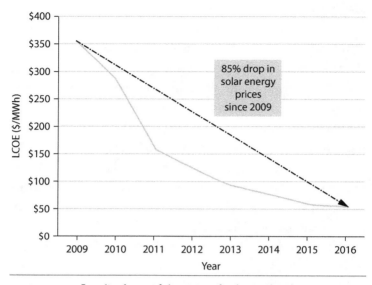

FIGURE 5.5 Levelized cost of electricity of utility-scale solar projects. (Adapted from Mike O'Boyle, "Wind and Solar Are Our Cheapest Electricity Sources—Now What Do We Do?," America's Power Plan, December 21, 2016. Data from Lazard's *Levelized Cost of Energy Analysis 10.0*)

85 percent in seven years due to the virtuous cycle of increased demand leading to increased production and application of the learning curve not only to the manufacturing of solar panels but also to the balance of systems and installation costs.[17] As LCOE falls, demand increases, and the cycle begins again, creating annual growth in the solar energy market of 68 percent every year over the past ten years.[18]

Early adopters of solar energy systems were primarily interested in the environmental benefits, the appeal of distributed generation, and the desire for energy independence. Today's consumer of solar energy is primarily motivated by cost savings. Homeowners, businesses, and utilities in many countries are finding that electricity generated from solar PV panels is now cost competitive with all other forms of electricity generation, which makes the logic for the switch to solar compelling. How did solar power become cost competitive with fossil fuels? Primarily by the application of the learning curve referred to as Swanson's law to relentlessly lower the LCOE of electricity generated with solar. A further explanation can be found in two essential developments in the early days of solar energy—government incentives and financial innovation—that provided financing to this capital-intensive sector.

FIGURE 6.1 Rio de Janeiro.
(Source: https://wikitravel.org/en/Rio_de_Janeiro)

6

FINANCING RENEWABLE ENERGY

WIND and solar projects generate electricity that is cost competitive with nuclear and fossil fuels, producing the cheapest energy in many places on the planet. But low cost is not the same as low investment. In fact, renewable energy projects require significant up-front capital investment for equipment and project construction. Thereafter, renewable energy projects require very little capital for ongoing operation, maintenance, and fuel since the wind and the sun are free.

This created a catch-22 in the early days of the wind and solar sectors because investors were unwilling to commit capital to a new and unfamiliar sector. But deprived of capital the wind and solar sectors could never reach scale and become cost competitive. To overcome this problem, the financing of renewable energy initially required government incentives designed to entice investors to provide capital to the nascent wind and solar sectors by reducing risks and improving returns. These incentives spurred growth in financing, which was then followed by financial innovations to meet rising demand for renewable energy from homeowners and small businesses.

GOVERNMENT INCENTIVES

Government policy makers are well aware of the potential for renewable energy to be a competitive source of electricity, given the benefits of the learning curve and the virtuous cycle of growth that leads to lower costs and further demand-driven growth. But the reverse is also true—when renewable energy costs more than traditional sources of electricity, demand will be weak, production will be low, and costs will never decline. This creates a hurdle for the entire renewable energy sector. Therefore, the role of government is to overcome that hurdle by providing incentives to ignite the virtuous cycle of increasing demand and decreasing costs. Governments do this with two types of incentives: command and control, and market mechanisms.

Command and control, which takes the form of direct subsidies or tax incentives, is the traditional approach for governmental support of an industrial sector. In the renewable energy sector, governments in Europe and most developing countries have implemented command and control subsidies using what are called feed-in tariffs. Feed-in tariffs provide renewable energy project developers with a long-term contract to sell all the electricity produced by the project at a fixed price that is higher than the prevailing market price for electricity. The difference between the fixed price of the feed-in tariff and the market price is the government subsidy.

For renewable energy project developers, the advantages of feed-in tariffs include not only the higher purchase price for electricity generated but also the long-term contract, which reduces risk for project investors. Feed-in tariffs are highly effective at motivating investors and project developers to rapidly finance and build renewable energy projects. However, feed-in tariffs can initially be costly for governments, as the subsidy is often very high.

Well-designed feed-in tariffs include a process that resets the tariff level lower for new projects as installation volumes increase.

In the United States, in lieu of a feed-in tariff the federal government has implemented tax incentives for renewable energy. Tax credits are a government subsidy provided to investors in renewable energy to compensate for the initial higher cost of building projects. Unfortunately, tax credits are a cumbersome and complex incentive, as the renewable energy project developer must find an investor willing to finance the project and one capable of utilizing the tax credit.

With command and control incentives, the government decides the subsidy rate, which it can lower over time as renewable energy becomes more competitive. The advantage of command and control government subsidies are that they are simple in design and targeted. The trade-off is that government subsidies are inefficient, as the government invariably sets the subsidy too high or too low, often with limited flexibility to change as the industry evolves. Command and control incentives are a relatively blunt, though effective, tool for incentivizing investment in renewable energy.

Harnessing the Market

Governments also use market mechanisms to incentivize renewable energy development. To create a market mechanism, the government establishes an incentive system and then allows the market participants, including renewable energy developers and utilities, to find the lowest-cost solution.

The simplest form of market mechanism is an auction in which governments issue a tender or request for new electricity generation and developers bid to supply electricity at the lowest price. Auctions incentivize renewable energy projects by providing

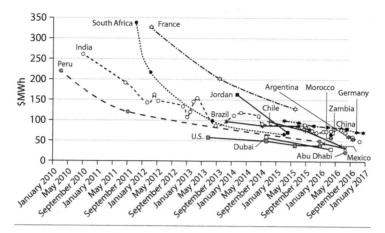

FIGURE 6.2 Evolution of auction prices for solar PV.
(Adapted from International Renewable Energy Agency,
Renewable Energy Auctions: Analysing 2016, http://www.irena.org
/publications/2017/Jun/Renewable-Energy-Auctions-Analysing-2016)

developers with a long-term government contract at a fixed price.
The advantages of auctions for governments are simplicity, trans-
parency, and lower electricity prices from competition among
developers. Auctions for renewable energy have been used in
forty-eight countries, including India, Mexico, and South Africa,
and countries that had previously offered feed-in tariffs, like
Germany, which are now offering competitive auctions.[1] The use
of auctions has simultaneously supported growth in renewables
while driving down costs, as demonstrated in figure 6.2 by the
evolution of auctions for solar power.

RPS Markets

A more sophisticated market mechanism, used primarily by
state governments in the United States, is a renewable portfolio

standard (RPS). In an RPS, the state government sets a minimum percentage of electricity to be generated from renewable sources and establishes a penalty for noncompliance. Electric utilities that are subject to a state RPS are required to use electricity from renewable energy sources to meet or exceed the minimum percentage. Utilities have the flexibility of buying electricity from independent project developers or other utilities, or investing in renewable energy projects. Flexibility is the primary advantage of an RPS, as it allows utilities to meet the target with the lowest-cost source of renewable power.

An RPS supports development of renewables by increasing demand for renewable energy, incentivizing utilities to sign contracts for renewable energy with project developers, and encouraging utilities to invest in the sector. In the United States, the RPS market mechanism is increasingly popular, with twenty-nine out of fifty states having adopted renewable portfolio standards to support the development of wind and solar power (figure 6.3).[2]

Carbon Markets

In 1992, representatives from 172 countries met in Rio de Janeiro, Brazil, for the first Earth Summit. Negotiators reached agreement on the Climate Change Convention, a global treaty with the objective to "stabilize greenhouse gas concentrations in the atmosphere at a level that would prevent dangerous anthropogenic interference with the climate system."[3] The treaty was signed by 154 countries, including the United States. The Climate Change Convention launched an annual global meeting to review and negotiate solutions to climate change. In 1997, at the annual meeting in Japan, negotiators produced the Kyoto Protocol, the world's

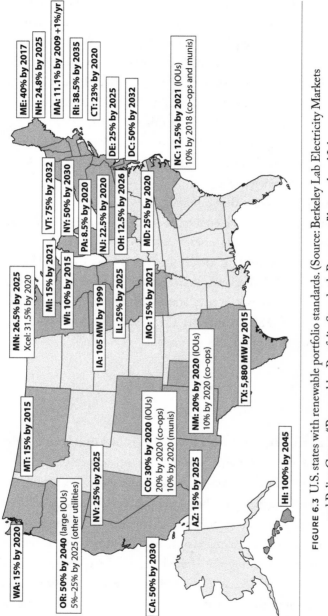

WA: 15% by 2020

OR: 50% by 2040 (large IOUs)
5%–25% by 2025 (other utilities)

CA: 50% by 2030

MT: 15% by 2015

NV: 25% by 2025

AZ: 15% by 2025

CO: 30% by 2020 (IOUs)
20% by 2020 (co-ops)
10% by 2020 (munis)

NM: 20% by 2020 (IOUs)
10% by 2020 (co-ops)

TX: 5,880 MW by 2015

HI: 100% by 2045

MN: 26.5% by 2025
Xcel: 31.5% by 2020

MI: 15% by 2021

WI: 10% by 2015

IA: 105 MW by 1999

IL: 25% by 2025

MO: 15% by 2021

VT: 75% by 2032

NY: 50% by 2030

PA: 8.5% by 2020

NJ: 22.5% by 2020

OH: 12.5% by 2026

MD: 25% by 2020

ME: 40% by 2017

NH: 24.8% by 2025

MA: 11.1% by 2009 +1%/yr

RI: 38.5% by 2035

CT: 23% by 2020

DE: 25% by 2025

DC: 50% by 2032

NC: 12.5% by 2021 (IOUs)
10% by 2018 (co-ops and munis)

FIGURE 6.3 U.S. states with renewable portfolio standards. (Source: Berkeley Lab Electricity Markets and Policy Group, "Renewables Portfolio Standards Resources," last updated July 21, 2017, https://emp.lbl.gov/projects/renewables-portfolio)

first agreement to reduce greenhouse gas emissions. It included the use of carbon markets to incentivize the transition from fossil fuels to renewable energy.

Carbon markets, also referred to as cap and trade, are designed to reduce greenhouse gas emissions at the lowest possible cost. The economic theory upon which carbon markets are based was first postulated by an obscure Canadian economist, John Dales, in 1968. Dales recognized that it is extremely difficult for any government to accurately tax emissions, as governments are unable to determine how much it will cost a business to reduce emissions—costs change constantly with new technologies, innovation, and market conditions. Businesses themselves are often uncertain of the cost of reducing pollution until they attempt to do so.

Not knowing the cost to reduce pollution makes it nearly impossible for a government or business to assign a price to emissions, and without a price there is no way to properly invest to lower pollution. Fortunately, Dales had an elegant solution, which was to place a cap on the overall level of pollution and assign rights allowing businesses to collectively pollute up to the cap but no more. Importantly, businesses would be allowed to trade their rights to pollute, creating a market. Dales wrote that "the virtues of the market mechanism are that no person, or agency, has to set the price—it is set by the competition among buyers and seller of rights."[4]

Dales's theory was not put to the test until 1990, when U.S. President George H. W. Bush supported an amendment to the Clean Air Act that allowed for the establishment of a cap-and-trade market to address emissions of sulfur dioxide from coal-fired utilities. When released in the atmosphere sulfur dioxide creates acid rain, poisoning fish and damaging trees, which by the 1990s was causing significant environmental harm, especially in the northeastern United States. The cap-and-trade program created under the Clean Air Act issued pollution rights to utilities which could then be traded. Utilities had the flexibility to reduce emissions below the cap or to buy rights from other utilities that reduced emissions at lower cost. Utilities that did neither faced a costly government fine.

The cap-and-trade program to reduce acid rain was a stunning success, reducing sulfur dioxide emissions by 50 percent at very low cost to business.[5] More importantly, the program proved that a market-based system was capable of capping emissions at a desired level, reducing pollution at the lowest overall cost, and incentivizing businesses to innovate low-emission technologies and solutions. The dramatic success of the U.S. acid rain program encouraged economists working on climate change to incorporate a similar system for reducing greenhouse gas emissions, which came to be called carbon markets.

Carbon markets play a role in the growth of renewable energy because wind and solar power do not emit any greenhouse gases. For utilities facing a cap on greenhouse gas emissions, wind and solar are an attractive option because they generate electricity without requiring any allowances to pollute. Utilities required to reduce greenhouse gas emissions below their cap can do so by purchasing electricity generated by renewable energy sources, buying renewable energy assets, or by investing in new renewable energy projects, in each case increasing the demand for renewable energy.

The Kyoto Protocol created a global carbon market, spurring investment in projects that reduced greenhouse gases. Investors in renewable energy projects received carbon credits that could be sold, generating additional revenue and improving returns on project investment. The country with the largest number of projects built in compliance with the Kyoto Protocol is China;

these include a significant number of wind projects that benefited from the carbon market, lowering the cost of wind power for Chinese consumers.[6]

Unfortunately, governments were unable to agree on an extension to the Kyoto Protocol before key parts of it expired in 2012, but carbon markets continue to be used as an incentive to develop renewable energy. European governments created the EU Emissions Trading System in 2004, providing an incentive for European utilities to switch from using fossil fuels to generate electricity to renewable wind and solar energy. California established a cap-and-trade program in 2012 as part of the state's plan to cost-effectively reduce greenhouse gas emissions by 80 percent from 1990 levels by 2050.[7] And in China, the government established carbon markets in 2013 in seven provinces, with plans to combine them into one national carbon market to create the world's largest cap-and-trade program.[8]

Carbon markets are designed by governments to reduce greenhouse gas emissions at the lowest cost possible, using the market to set the price of emissions and thereby guide investment decisions. This can create a very efficient allocation of capital to low-emission energy generation technologies, including solar and wind power. However, establishing a carbon market requires governments to enact legislation that caps greenhouse gas emissions; regrettably, the political will to do so is often lacking.

Government's Declining Role

Government incentives play an essential role in the growth of renewable energy via command and control subsidies (feed-in tariffs and tax credits) and market mechanisms (auctions, renewable portfolio standards, and carbon markets). These incentives reduce risk and increase returns for renewable energy investors. The combination of lower risk and higher returns has incentivized investors to provide capital to grow the wind and solar sectors, creating a virtuous cycle of lower costs, higher demand, and increased production that takes advantage of the learning curve to further lower costs and start the cycle all over again.

Importantly, most government incentives for renewable energy are designed to decrease over time as the LCOE of electricity from wind and solar declines until reaching a point referred to as parity, when these technologies are cost competitive without any additional incentives. (Parity is addressed in chapter 9.) Government support has provided a critical role in the growth of renewables, drawing private capital into financing the sector. Financing renewable energy began with traditional structures like project finance, followed by financial innovation to meet the growing need for investment capital.

Financing Renewable Energy

Recall the three categories of renewable energy projects: residential for homeowners, commercial for building owners, and utility scale for large-scale electricity generation for the grid. These distinctions are also helpful for understanding the financing of renewable energy. Residential and commercial scale projects were initially very difficult to finance because of their small scale. Solving that problem required financial innovation, discussed later in this chapter. But financing utility-scale wind and solar projects was reasonably straightforward, using a traditional investment structure called project finance.

FINANCING UTILITY-SCALE RENEWABLE ENERGY PROJECTS

Project finance is the financing of infrastructure based on the expectation of stable, long-term cash flows. The history of project finance extends back many hundreds of years. The English Crown is believed to have used project finance in 1299 to develop the silver mines in Devon, borrowing the capital from an Italian merchant bank in return for a portion of the output from the mines.[9] In the twentieth century, project finance was used to finance toll roads, canals, railways, and power projects globally. These projects all share similar traits—they are large, long lasting, and capable of generating stable cash flows to repay investors. Utility-scale wind and solar projects also share these attributes.

The key to raising project finance capital is to demonstrate that wind and solar projects can generate highly predictable cash flows over many years. Fortunately, forecasting cash flows from wind and solar projects is straightforward:

1. Electricity generation from renewable projects is predictable; day to day the wind and the sun are highly variable, but over a year the average wind speed or days of sunshine are highly consistent, and

2. the electricity generated by wind farm or solar projects is presold at a fixed price for the life of the project using a power purchase agreement (PPA), a contract between the project owner and utility buying the electricity. The fixed price is determined in negotiation between the project owner and the utility, or by the government if a feed-in tariff is available. Therefore,

3. revenue from the project is simply the forecasted electricity generation multiplied by the PPA price. The net cash flow to the project investors is the project revenue less operating and maintenance costs.

Wind and solar projects generate electricity for twenty-five years or more. With zero input costs, very low operating and maintenance costs, and fixed price contracts to sell all electricity generated, forecasting long-term cash flows is relatively easy. This makes renewable energy projects ideal for raising capital using the project-finance structure, as lenders can be confident they will be repaid over the life of the project. Stable cash flows mean lower risk, encouraging lenders to provide funds at attractive interest rates, thereby reducing the project's cost of capital.

A low cost of capital is important if wind and solar are to compete against other forms of electricity generation. Since renewable energy projects operate for twenty-five years or longer, and because almost all the capital is required upfront for project construction, the LCOE is very sensitive to the cost of capital required to finance the project. The investment bank Lazard has calculated the impact that decreasing the cost of capital has on LCOE. For example, lowering the cost of capital from 9.2 percent to 5.4 percent cuts the LCOE of a solar project by more than a third.[10] Investment capital must be sourced at low rates to keep the LCOE on solar and wind projects competitive with other sources of electricity generation. Fortunately, the project-finance structure does exactly that.

Project finance has become the most popular form of financing for large utility-scale wind and solar projects, raising hundreds of billions of dollars of capital. And that capital is at very low interest rates. In 2016, wind developers in France used project

FIGURE 6.4 The London Array, the world's largest offshore wind farm, generating power for five hundred thousand homes. It was funded using project finance from a consortium of banks. (Photo courtesy of Scira Offshore Energy Ltd. / Statkraft via Flickr)

finance to raise fifteen-year loans at an incredibly low 2.5 percent, down from an already low 4 percent in 2014.[11] Loan sizes have also increased as lenders have become more comfortable with the renewable energy sector—an offshore wind farm near the coast of England is raising over $5 billion for a single project.[12]

Project finance, however, requires negotiation and documentation for each individual project. This process is time consuming, taking many months or even years, and expensive, requiring sophisticated legal and financial expertise. Therefore, project finance is rarely used for projects of less than $100 million in capital costs. This makes it an ideal solution to source funds for large utility-scale renewable energy projects. But it does not provide financing for smaller wind and solar projects. That requires financial innovation.

FINANCING RESIDENTIAL AND COMMERCIAL RENEWABLE ENERGY PROJECTS

Homeowners and businesses have been very reluctant to spend the large amount of upfront capital necessary to place a solar power system on the roof of their homes or buildings, even though they understand that solar energy can provide them with cost competitive electricity. And there are potential risks. What if the system fails to generate electricity as predicted? What if the homeowner moves? Who will service or repair the system? In 2004, an innovative financial mechanism called solar leasing was created to address these concerns for households and owners of small buildings.

Solar Leasing and Pay-As-You-Go

In a solar lease, the home or building owner signs a contract with a solar project developer agreeing that all electricity generated from the solar panels will be used in the house or building at a fixed price for the life of the project. In return for the contract by the building owner to buy the electricity generated at a fixed price, the developer constructs the solar project and performs all maintenance and repairs for the life of the project. Most importantly, the developer finances the upfront capital cost of the solar project.

The importance of solar leasing to the growth of solar energy cannot be overstated. In 2009, *Scientific American* placed it among twenty "world-changing ideas."[13] Solar leasing allowed homeowners to put solar panels on their roof without having to put any money down. And it allowed them to use solar energy without taking any risk. Unsurprisingly, this financial innovation fueled a dramatic increase in solar energy systems among homeowners. In the United States, the percentage of new residential solar systems that were leased or otherwise owned by third parties grew to nearly three-quarters of new installations by 2014.[14] The acceptance of solar leasing was the primary reason the residential solar market in the United States experienced growth greater than 50 percent every year from 2011 through 2014.[15]

A variation of the solar lease model, called pay-as-you-go (PAYG), has been designed for households in developing countries. The PAYG model allows customers to purchase a solar system over time. After making an initial deposit, small payments are then made over months or years, usually via a mobile phone payment system. Once the user pays off the system, the solar power is free in perpetuity. PAYG alleviates the high up-front costs that had previously created a barrier to solar adoption in developing countries.

FIGURE 6.5 The Musau family in Athi River, Kenya, enjoying pay-as-you-go power and appliances from M-KOPA Solar. (Photo © Allan Gichigi/M-KOPA Solar)

PAYG has been used to install solar systems in dozens of developing countries, and it is growing rapidly. PAYG financing models are helping families in poor countries access lights and phone chargers, which allow them to save money on charging their phones and allows their children to do schoolwork after dark.[16] Not only do solar lights save money, they eliminate the danger of fire associated with kerosene lamps. PAYG makes economic and practical sense for consumers, makes financial sense for companies and investors, and reduces unhealthy emissions and safety concerns arising from the burning of wood and kerosene. In recognition of the potential of this model to profitably bring electricity to the 1.2 billion people living outside the electric grid globally, investment capital has flowed into companies offering PAYG solar systems in developing countries.

INSTITUTIONAL INVESTORS

The dramatic growth in demand for solar power brought about by solar leasing and the pay-as-you-go financing model created a need for billions of dollars of low-cost capital to fund those businesses, as did the growth in utility scale projects funded with project finance. Raising large volumes of capital required the participation of institutional investors, entities that include pension funds, money managers, and insurance companies.

Institutional investors were initially reluctant to finance wind and solar projects due to their lack of experience in the sector. As renewable energy became increasingly popular, investors were drawn to the stable, long-term cash flows generated by wind and solar. By 2011 investors were pouring over $200 billion per year into renewables, almost entirely into wind and solar.[17] *Bloomberg* reported that pension funds and other institutional investors were investing greater sums in wind and solar projects because "the industry is considered a safe alternative to traditional securities such as government bonds."[18] As institutional investors became increasingly comfortable with the sector, the required returns on investment declined. And the lower the return required by investors became, the lower the LCOE or cost of generating electricity from wind and solar became in turn. A lower LCOE, of course, makes renewable energy more competitive with other sources of electricity generation such as coal or natural gas, increasing demand for renewables and creating a virtuous cycle, due not only to lower project costs but also to lower financing costs.

HURDLES AHEAD

Wind and solar power, having become cost competitive sources of electricity, are on the cusp of the next great energy transition. Demand for renewable energy, initiated with government subsidies, ignited a learning curve that drove down costs, led to further demand, and created a virtuous cycle of cost reductions. And innovative financing reduced the cost of capital for wind and solar projects, further improving the competitiveness of renewable energy.

During this early growth phase the wind and solar sectors have confronted many hurdles, from NIMBYism to regulatory barriers. Many of these barriers have been addressed through increasingly lower costs and competitiveness; the cheaper and more competitive that wind and solar become, the harder it is for regulators or competitors to limit further growth. There are, however, two hurdles that stand in the way of a full transition from fossil fuels to renewable energy.

The first hurdle results from the structure of the existing electric utility system. The grid was never designed for generation of electricity at countless locations, which occurs when homeowners and businesses install solar and wind power. In most countries, regulated utilities generate electricity at large centralized facilities, transmit power across an electrical grid, and deliver the electricity to homes and businesses. Utilities and transmission systems are most often regulated monopolies or oligopolies, as it is impractical to build redundant infrastructure, especially when it is very costly to construct and maintain those facilities. The generation and transmission system of electric utilities is nearly unchanged since Thomas Edison opened the Pearl Street Station in 1882. What has changed is the potential for distributed energy, specifically wind and solar power, to disrupt the traditional utility model.

Renewable energy can be installed behind the meter, allowing homeowners and businesses to generate their own electricity. And if the homeowner's solar panels generate excess electricity, it can be sent back to the grid for use by the utility, in return for which the homeowner receives a payment or credit on his or her utility bill. In this way, distributed generation can create a tremendous challenge for utilities, both taking away customer demand and forcing utilities to purchase excess power production. Recent disputes between utilities, regulators, and homeowners over net metering are an unsurprising outcome of distributed generation, and they will need to be resolved to ensure continued growth in renewables for households and businesses.

A modern grid capable of economically and equitably serving all sources of electricity generation, including distributed solar and wind projects, is a prerequisite to the energy transition. Technologically, that is already feasible, but updating regulations will take more time. Unfortunately, redesign of the regulatory framework overseeing the electrical grid faces political and bureaucratic challenges from incumbent utilities unwilling or unable to adjust to a rapidly changing industry. But the popularity of renewable energy among consumers is forcing regulators to seek a solution to this first hurdle to continued growth in wind and solar.

The second and more significant hurdle is the challenge of intermittency. Electricity generated by wind turbines and solar panels cannot be dispatched at will, nor can electricity be cheaply stored for later use. Solving the intermittency problem is the final hurdle before wind and solar can become the dominant global source of electricity generation. In order to understand the solution to intermittency, a step back is in order. The next chapter returns to the topic of energy transitions, examining a parallel energy transition that took place in the twentieth century, from feeding horses to powering automobiles, ships, and aircraft. Understanding the arc of the energy transition in transport leads to a solution to the intermittency problem and the eventual dominance of renewable wind and solar energy.

FIGURE 7.1 The Benz Patent-Motorwagen.
(Source: Wikimedia Commons)

7

ENERGY TRANSITIONS

Oats to Oil

ARLY humans invented the wheel some four thousand years ago, a development roughly coinciding with the domestication of the horse. And for the following four millennia, overland transport options were limited to the horse and other domesticated work animals. Horses became ubiquitous in the United States and Europe, consuming vast amounts of hay, oats, and other feed and depositing a nearly equal quantity of waste. By 1900, London was home to over eleven thousand horse-drawn cabs and thousands of horse-drawn buses, requiring more than fifty thousand horses. New York City residents at the time shared the streets with one hundred thousand horses.[1]

Feeding and caring for all that transportation was costly. The source of power for animals, of course, is feed, the unit of power of which is calories. A workhorse requires an estimated twenty-five thousand calories per day, equivalent to twenty pounds of oats.[2] Horse owners incurred a significant expense in supplying all that animal feed. And the resulting mess was a serious problem for everyone else; in 1894, the situation in London became so acute that the *Times* predicted that "in fifty years, every street in London will be buried under nine feet of manure."[3]

Early inventors contemplated a successor to the horse. In 1509 Leonardo da Vinci described a theoretical compressionless engine, a precursor to the modern internal combustion engine.[4] But it was not until the German inventor Carl Benz invented his "horseless carriage" in 1885 that a workable solution emerged. Benz chose the internal combustion engine, powered by gasoline, to propel his invention, an unusual choice at the time.[5]

When Benz designed the first automobile, the internal combustion engine already existed but was not widely used. Motorized transport at that time was exclusively by train or boat, technologies that used coal-powered steam engines. Benz was attracted to the internal combustion engine because it is more efficient than a steam engine, meaning that a greater percentage of the energy in the fuel would be converted to mechanical power. Another important feature of the internal combustion engine is its portability, which makes it perfectly suited not only for automobiles but for many forms of transportation and farm equipment.

In 1885, the primary fuel in the global economy was coal, which had earlier replaced wood, as discussed in chapter 2. Coal was a cheap and readily available source of power, and compared to wood it is relatively energy dense. However, coal is a solid, and the internal combustion engine requires fuel in a liquid or gaseous state to operate. Gasoline is produced by refining crude oil, the resultant product of which is nearly twice as energy dense as coal. High energy density meant that Benz's automobile required less fuel per mile than a coal-powered engine. And given that gasoline is a fluid, it could be stored in a simple fuel tank.

Within two years, Benz began commercial production of his "horseless carriage," the first gasoline-powered vehicle. Remarkably, the principles behind the design of the internal combustion engine have not changed significantly since that time. Mass production of the Ford Model T in 1908 led to a boom in automobile

FIGURE 7.2 The first oil gusher, Spindletop, Texas (1901).
(Source: Wikimedia Commons)

sales and greatly increased the demand for gasoline. Consumption of gasoline tripled between 1919 and 1929,[6] closely tracking the rise in ownership of motorcars. The energy transition from oats to oil had begun. As automobile sales grew, so too did the demand for oil, and drilling for oil boomed. In 1901, a well drilled at Spindletop, Texas, blew, gushing out one hundred thousand barrels of oil per day and establishing the modern oil industry.

FROM COAL- TO OIL-POWERED NAVIES

In 1911, coal was the main source of power for Great Britain's Royal Navy. But it was not without drawbacks. During World War I, Britain realized that it could not achieve the speed necessary to outmaneuver its opponents without oil-fueled battleships, marking the beginning of the transition of its fleet. The energy density of oil over coal meant that ships' boiler sizes could be reduced and the miles traveled before refueling extended. There was a tactical benefit too: the reduced smoke output from oil over coal meant that ships were less visible to enemy lookouts.[7]

As is often the case with energy transitions, the availability of raw materials was an important practical and logistical consideration, as were the attitudes of those in power and of end users. In Britain's case, the country had large coal supplies but did not have oil reserves. The British cleverly addressed this challenge by taking a 51-percent stake in the Anglo-Persian Oil Company and negotiating a twenty-year contract for the provision of oil to the Royal Navy. As Winston Churchill observed, "The advantages conferred by liquid fuel were inestimable." But, he went on, "To change the foundation of the navy from British coal to foreign oil was a formidable decision in it-self."[8]

In the United States, the last coal-fired battleship was commissioned in 1914.[9] Range limitations associated with the use of coal became apparent during the Spanish-American War, during which the U.S. fleets ventured further from shore than ever before. As in the case of the British Navy, this transition allowed the U.S. Navy to realize the benefits of oil's energy density and the relative advantages of internal combustion engines over steam engines.

A COMPLETE TRANSITION: OATS TO OIL

By the mid-1950s, petroleum products were supplying more than 90 percent of the fuel for transportation in the United States.[10] Soon thereafter nearly all U.S. farms had replaced animals with tractors.[11] One might assume that the nearly complete transition from oats to oil was swift and universal. But nearly thirty years after Benz's first commercial application, the U.S. Navy was still commissioning coal-fired ships despite the obvious advantages of gasoline and internal combustion engines. Steam-powered locomotives were still in use in Europe and the United States in the 1960s.

As with all energy transitions, the move from animal feed to oil was slow and its outcome difficult to predict at the time, yet economic forces eventually overcame all resistance to change. Oil now fuels 94 percent of transportation in the United States.[12] The measurement of engine power in "horsepower" illustrates just how linked our modern machines are to their animal predecessors.

LESSONS FROM THE PAST

The transition from oats and coal to oil seems inevitable in retrospect. Automobiles are clearly faster and more powerful than horses. But the desirability of the transition was less evident at the time. Horses were familiar to all as the means for local transportation. Coal was the fuel of choice for motorized transport on railroads and ships. And electricity was the most exciting innovation of that era, a near-magical form of energy. The internal combustion engine, powered by gasoline, was not the only option. The next chapter will revisit the decision by Benz and other automobile pioneers to choose the internal combustion engine and will examine the quixotic rise of electric vehicles.

FIGURE 8.1 Elon Musk, CEO of Tesla.
(Source: Wikimedia Commons)

8

THE RISE OF ELECTRIC VEHICLES

ENRY Ford and other early automakers believed "there could be nothing new and worthwhile that did not run by electricity. It was to be the universal power."[1] Despite the success of Benz's gasoline-powered automobile, by 1900 one-third of all vehicles on American roads were electric cars. New York City had a fleet of electric taxis, and the first-ever speeding ticket was issued to an electric taxi in Manhattan.[2] Even sports car legend Ferdinand Porsche developed an electric car.[3] The popularity of electric cars was self-evident; they were quieter, cleaner, and easier to operate than vehicles powered by internal combustion engines. Electric cars did not require a hand crank to start and did not emit noxious-smelling exhaust. Convinced by its many advantages, Thomas Edison and Henry Ford created a partnership in 1914 to design a low-cost electric car.

Edison and Ford well understood that electric vehicles (EVs) have several advantages over vehicles powered by internal combustion engines (ICEs). Foremost among these is that EVs convert energy into motion much more efficiently than ICEs. ICEs burn gasoline to create heat, a process called combustion; the heat is then converted into mechanical energy to propel the vehicle. This process is inefficient, as only 17–21 percent of the energy stored in gasoline is converted to mechanical power. Electric motors, on the other hand, are highly efficient. By directly converting electrical energy into mechanical energy, an EV converts 59–62 percent of the electrical energy into powering the wheels.[4] Because the motor is more efficient, an electric vehicle requires one-third of the energy of a gasoline-powered vehicle.

But the Achilles' heel of electric cars is the battery. All vehicles require a medium to store the energy that propels them. Electric vehicles store energy in batteries, while vehicles powered by internal combustion engines store energy in gasoline. Recall once again the importance of energy density, defined as the amount of energy that can be stored per unit volume or mass. Batteries at the dawn of the automobile age were large and stored relatively little electricity, so energy density was low. Gasoline, by comparison, is energy dense. The same volume of gasoline can store more than one hundred times the energy in a lead-acid battery, the most common type of battery in Edison's day. This gave ICEs a distinct advantage over EVs in the early 1900s because gasoline is a much better way to store energy than the best batteries of the time; this advantage more than made up for the relative inefficiency of ICEs in converting that energy to mechanical power.

Edison and Ford were aware of the storage problem in EVs, and they tackled it by working to invent a battery with greater energy density. Henry Ford, quoted in the *New York Times* in January 1914, asserted: "The problem so far has been to build a storage battery of light weight which would operate for long distances without recharging. Mr. Edison has been experimenting with such a battery for some time."[5]

It did not work. Thomas Edison was the greatest inventor of his time, but he failed in his quest to build a better battery, ending the joint venture with Ford, and along with it the potential of EVs.[6] It was nearly one hundred years before electric vehicles would become commercially viable.

INNOVATION IN THE AUTO INDUSTRY

The oil crisis of the 1970s spurred automakers to reconsider electric vehicles as an alternative to gasoline-powered internal combustion engines. Engineers revisited the dilemma Edison and Ford faced and set out to design a battery that was lightweight yet energy dense, capable of storing enough energy to power a modern car. Again, they failed.

In 1990, California enacted a program to promote the use of zero-emission vehicles, which created a strong incentive for automakers to design and produce electric cars. The draw of selling cars in California convinced the major automakers to once again invest research and development dollars to create a successful electric vehicle. General Motors introduced the first mass-produced all-electric car, the EV1, in 1996, with a lead-acid battery weighing 1,175 pounds and a range of 60 miles per charge.[7] But the vehicle was expensive to produce and the limited driving range was not appealing to consumers. GM canceled the EV1 program after only a few years and destroyed the cars it had built, claiming the vehicle was not commercially viable.

Regrettably for GM, the company's ill-fated foray into EVs inspired the documentary "Who Killed the Electric Car?," which suggested a conspiracy between the auto and oil companies to prevent the development of electric vehicles. While a conspiracy is unlikely—GM vehemently denied it—the company's decision to destroy every EV1 produced (by crushing them) was surely one of the more poorly considered plans in business history from a public relations perspective. More importantly, despite significant research and development, GM, Chrysler, Ford, Toyota, Honda, and other leading automotive companies were incapable of designing an electric car that would travel as far or as fast as gasoline-powered vehicles. Consumers were for the most part unimpressed with the electric cars that were offered; every electric model designed during this period was eventually withdrawn from the market.

In 2003, a new automobile company was formed, Tesla Inc., named after Nikola Tesla, inventor of the AC electric induction motor. Elon Musk, Tesla's CEO, announced the following "Secret Tesla Motors Master Plan (just between you and me)":

Build sports car
Use that money to build an affordable car
Use *that* money to build an even more affordable car.[8]

Musk's announcement was tongue-in-cheek, but Tesla's business plan was very serious and very innovative in several ways. First, Tesla targeted the high end of the car market, the sports car and luxury sector, for its initial products. Previously, automobile companies had always focused on the low-end or economy market to sell EVs. Second, Tesla focused on vehicle performance as the primary selling feature, as opposed to environmental benefits. Third, and most importantly, Tesla recognized that the solution to Edison's battery challenge was not to invent a new kind of car battery but to take advantage of the extraordinary progress already achieved in batteries for an entirely different market—consumer electronics and mobile phones.

Counterintuitively, the decision by Tesla to use existing batteries instead of developing a new, more energy-dense battery was highly innovative. Tesla recognized that the fastest developments in batteries were occurring in the consumer products markets, as demand for laptop computers and mobile phones inspired the world's leading technology companies to design and manufacture increasingly energy-dense, cost-competitive batteries.

Lithium-ion (LiOn) batteries were commercialized fewer than twenty-five years ago by Sony,[9] coinciding with the advent of consumer electrics (the most prominent of which were dubbed "the three Cs"—computers, cameras, and communication devices). Given Sony's leadership in these markets, it made sense for the company to be in the battery business as well. The relatively high energy density of lithium-ion batteries made them appealing for many consumer products.

Instead of developing a new, special-purpose car battery, the engineers at Tesla designed a battery pack composed of seven thousand lithium-ion batteries. The battery pack took advantage of the best characteristics of lithium-ion batteries—it was relatively small, lightweight, and energy dense at a reasonable cost. Rather than spend time and money inventing a new battery, the company's engineers focused on developing the software to control the battery pack, and they designed a proprietary powertrain connecting the batteries to an all-electric engine.

The first car that Tesla launched, the Roadster, was expensive, attractive, and fast. Electric motors produce significantly more torque—the rotational force from the motor to power the wheels—than an internal combustion engine, allowing for greater acceleration. Tesla's Roadster accelerated from zero to 60 mph in a remarkable 3.9 seconds. In 2008 *Motor Trend* reviewed the car, concluding that it is "profoundly humbling to just about any rumbling Ferrari or Porsche that makes the mistake of pulling up next to a silent, 105 mpg[10] Tesla Roadster at a stoplight."[11]

OVERCOMING CHALLENGES

The Roadster proved that an electric car could perform competitively against high-end sports cars. Tesla had successfully completed the first phase of its master plan. The second phase, the building of a competitive high-end luxury car, was significantly more ambitious, requiring additional innovation. Tesla's next car, named the Model S, was a sedan designed to compete with the BMW 5 Series and similar luxury cars. To be successful, Tesla had to overcome three challenges simultaneously: range anxiety, performance, and cost.

"Range anxiety" is a driver's fear that the battery in an electric car will run out of power before reaching the destination or a place to recharge. Prior to Tesla, commercial EVs had never exceeded a range of 60 miles, while gasoline-powered cars typically average well over 300 miles on a single tank and can be conveniently refueled in five minutes at any gas station. In theory, electric cars can be recharged using a regular electrical outlet, but in practice recharging from a 110-volt household outlet requires several hours or longer. Tesla tackled range anxiety in two ways. They made the battery pack significantly larger in the Model S, expanding the range between charges to between 210 and 300 miles, and they began installing "superchargers," high voltage charge stations, across the country. Superchargers allowed Tesla drivers to recharge their batteries to a range of 150 miles in approximately thirty minutes.

The expanded battery pack in the Model S was heavy; this is unsurprising given the relatively low energy density of batteries. The additional weight risked affecting the vehicle's performance. Tesla addressed this by placing the battery pack on the floor of the vehicle, between the axles, which improved handling by lowering the car's center of gravity. The company also increased the engine's power.[12] The result was astonishing. The Model S could accelerate faster than its competitors (it went zero to 60 mph in fewer than four seconds) and handled better. *Motor Trend* magazine rated it "Car of the Year" in 2013.

Cost remained the single biggest challenge for Tesla. The battery pack in the Model S was estimated to cost $15,000–$18,000, accounting for 25–30 percent of the total manufacturing cost of the car.[13] Government incentives offset some of these additional costs. In the United States, the federal government has allowed tax credits of up to $7,500 for the purchase of all-electric and plug-in hybrid vehicles. Tellingly, the size of the tax credit is dependent upon the capacity of the battery in the vehicle (the bigger the battery, the bigger the tax rebate).[14] Beyond the federal government's incentive, state and local jurisdictions offer a range of monetary and non-monetary incentives to electric vehicle owners. This goes beyond tax credits to include rebates or vouchers, reductions in vehicle registration fees, subsidized charging rates, and high-occupancy vehicle lane exemptions, among other incentives.[15]

In China, where electric vehicle sales have soared,[16] the national government has contributed to the cars' affordability by offering consumers exemptions of up to $8,500 in taxes. It also allows city governments to offer additional incentives of up to 50 percent of the value of the national incentives. Other countries, such as the Netherlands, have more complicated financial schemes that aim to incentivize consumers to purchase zero-emission vehicles.

In the Netherlands, registration taxation for vehicles is based on a vehicle's CO_2 emissions. These examples illustrate the power of a tool governments have long used to shape consumer behavior: taxation policy.

Government incentives, however, are generally designed to decrease over time or expire altogether. For example, in Denmark the government has begun phasing in registration taxes for electric vehicles from which they were once exempt. What started as a tax of only one-fifth of the full rate in 2016 will by 2022 reach the full taxation rate. To compete with gasoline-powered cars, Tesla had to source lower-cost batteries. Musk knew, however, that the prices of lithium-ion batteries would decline over time even if Tesla did nothing.

The key to Musk's strategy was that Tesla did not have to find ways to reduce battery costs. Instead, the company could count on battery manufacturers to do so in response to competitive pressures in the consumer electronics and mobile phone markets.

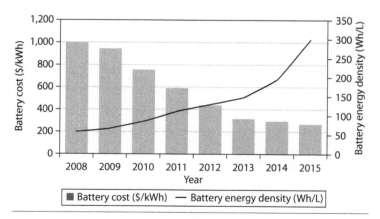

FIGURE 8.2 Evolution of battery energy density and cost.
(Adapted from International Energy Agency, *Global EV Outlook 2016*,
https://www.iea.org/publications/freepublications/publication
/Global_EV_Outlook_2016.pdf)

Early on, Tesla recognized the importance of the learning curve to batteries, specifically to lithium-ion batteries, which were widely used in many industries. Musk's master plan for Tesla allowed the company to take advantage of the learning curve of batteries used in mobile phones and other consumer electronics products.

Research on lithium-ion batteries has demonstrated a learning curve of 17 percent, meaning that costs decline by that amount each time the cumulative production of batteries doubles.[17] This was happening every three years due to booming demand from mobile phone and laptop manufacturing companies.[18] Tesla was positioned to take advantage of and benefit from the growth in those other markets, thereby addressing the single biggest cost in building the Model S and subsequent vehicles. It worked. The benefits to Tesla are apparent in figure 8.2; lithium-ion battery costs declined 73 percent, from $1,000 per kWh to $270 per kWh, in only seven years.

INNOVATION BEYOND THE BATTERY

Musk believed that Tesla's all-electric cars created the opportunity for further innovation in the automobile industry. Electric vehicles are much simpler than vehicles powered by internal combustion engines because they have fewer moving parts. With their simple design, EVs require fewer repairs and less servicing, a significant value to consumers. Analysis by JPMorgan found that "EVs have 20 moving parts compared to as many as 2,000 in an ICE, dramatically reducing service costs and increasing the longevity of the vehicle" and concluded that the running costs for an electric vehicle can be 90 percent lower than for a gasoline-powered one.[19]

This design advantage led Tesla to make the innovative decision to sell its vehicles directly to consumers through a network of company-owned stores, instead of the franchise dealerships used by other mass-market automobile companies. Musk believed that franchise dealers have a fundamental conflict of interest, as the typical dealer earns significantly more on replacement parts and service than from new car sales.[20] EVs like the Model S, however, are unlikely to require as many repairs as automobiles with ICEs, potentially upending the business model of the franchise dealer. Tesla's direct-distribution model is designed to minimize service costs for consumers over the life of the vehicle, thereby further improving the cost competitiveness of EVs in comparison with ICEs.

RESULTS

The Model S was an immediate success. Launched in 2012, it quickly became the top-selling all-electric car globally. In 2015, *Car and Driver* named the Model S the "Car of the Century." Consumer demand was so strong that Tesla started a waiting list. The United States accounted for the majority of sales, with strong results throughout Europe and even in China.

The Model S had a higher initial cost than comparable vehicles but lower running costs. Analysts at investment bank Credit Suisse estimated that the average owner of a Model S would spend $34 per month on fuel costs, compared with up to $175 per month for the equivalent gasoline-powered, midsized luxury sedan, and significantly lower servicing costs due to the simplicity of the electric motor.[21]

Customers raved about the vehicle. One clever Tesla owner in Sweden even wrote a parody, imagining a situation in which he test drove a gasoline-powered car for the first time. Unsurprisingly, the imaginary test drive did not go well, ending with the driver feeling sad for "our poor fellow commuters, who still had to put up with their gasoline cars."[22]

Tesla became the first American automobile company since World War II to successfully enter the mass market and the first to offer exclusively EVs. But the success of the Model S created a problem for Musk and his team. Tesla had taken advantage of the learning curve of lithium-ion batteries for laptops and mobile phones, following growth in those sectors to lower costs. To execute on the third part of Tesla's Secret Master Plan, to "build an even more affordable car," the company had to further reduce the cost of lithium-ion batteries. However, the demand for lithium-ion batteries to supply Tesla's vehicles was starting to outstrip demand from laptops and mobile phones. It was estimated that production of the Tesla Model S in 2014 would account for 40 percent of the demand for all lithium-ion batteries manufactured globally.[23] This prompted Musk to try another innovation, the Gigafactory, with the objective of bringing the learning curve for lithium-ion batteries in house.

BIRTH OF THE GIGAFACTORY

In 2014, Tesla broke ground in Sparks, Nevada, on what it calls the Gigafactory, the world's biggest lithium-ion battery factory. It is difficult to overstate its size. Upon completion, the Gigafactory will be the biggest building in the world, with the capacity to manufacture 35 gigawatt hours of lithium-ion batteries per year, more than all other lithium-ion factories combined.[24] That is enough batteries to build over four hundred thousand all-electric Tesla cars annually.

Most importantly, the Gigafactory was designed to lower the price of Tesla's battery packs by 30 percent by 2020, taking advantage of the learning curve.[25] This will create a virtuous circle for Tesla. Lower battery prices make Tesla's cars more competitive, increasing demand for them, increasing the production of batteries, decreasing cost of manufacturing, and thus again driving up demand.

The final step in Tesla's original master plan, "build an even more affordable car," was realized in 2016 with the release of the Model 3. Within a week, Tesla received 325,000 preorders, with a $1,000 deposit on each order. The Model 3 is a mid-sized, all-electric sedan with a range of 215 miles, designed to compete with the BMW 3 Series and similar vehicles. To meet demand, Musk announced that Tesla would start developing "probably four" new Gigafactories.[26]

IMITATION AND COMPETITION

The surprising success of Tesla's electric vehicles spurred incumbent automobile companies to design and launch their own EVs. Toyota had already experienced tremendous success with the Prius, a hybrid vehicle that combined an electric motor and a small ICE. Sales of the Prius, launched in 1997, reached nearly two hundred thousand cars in 2007.[27] Several other automobile companies launched hybrid vehicles similar in design to the Prius. All-electric vehicles, however, were viewed by mass-market car companies as unlikely to achieve commercial success. The complete failure by car companies in the 1990s to build a successful EV for the California market was a discouraging precedent, and the challenges faced by Tesla, including range anxiety, performance, and battery cost, convinced most manufacturers to wait for the

EV market to develop. But one incumbent car company, Nissan, saw the potential for using lithium-ion batteries, although the strategy the company took differed markedly from that of Tesla.

In 2007 Renault, the parent company of Nissan, initiated the most ambitious EV project of any mass-market automobile company. Nissan invested $5.6 billion to develop the Leaf, the first all-electric, mid-sized family sedan designed for mainstream drivers. Like Tesla, Nissan made the decision to use lithium-ion batteries. The Leaf's battery pack was smaller, however, allowing for a driving range of just 100 miles. While Tesla had targeted the sports and luxury end of the car market with high-performance vehicles, Nissan went after the mid-market with a car that focused on energy savings and environmental benefits.

Electric vehicles are significantly cheaper to run than gasoline-powered vehicles for two important reasons: electric motors are far more efficient at converting energy into forward motion, and electricity is a cheaper source of power than gasoline. In the United States, the average gasoline-powered automobile gets 22 mpg, which at $2.30 per gallon works out to $0.11 per mile. The average electric vehicle gets 4 miles per kWh, which at an average consumer cost of $0.13 per kWh[28] works out to $0.03 per mile, a rate that is 73 percent less costly than that of gasoline.[29] Another way to look at operating costs is in miles per dollar. Using these same figures, an electric vehicle can travel over 33 miles on $1, while a gasoline-fueled vehicle travels only 9 miles. With these advantages in operating cost, Nissan targeted its new EV at economy-minded consumers. The Leaf was launched in late 2010 to excellent reviews and became the world's best-selling EV in 2012.[30]

The success of Tesla's Model S at the high end of the market, and the success of the Nissan Leaf at the low end, gave automakers a glimmer of hope that EVs could be commercially successful.

There remained, however, several daunting hurdles to growth in the EV sector.

HURDLES TO GROWTH

Range anxiety, the fear that the batteries in an EV will run out of energy before a driver reaches his or her destination, has troubled the industry from the beginning. In fact, range anxiety by consumers is almost entirely irrational; analysis of driving habits finds that 95 percent of journeys taken by American drivers are under 30 miles, and 98 percent are under 50 miles in length.[31] Even the smallest EVs are capable of that range without recharging. The obvious solution to range anxiety is to increase the size of the battery in an EV; however, a larger battery adds significantly to the cost of an EV, making it uncompetitive on price with ICEs. The cost of batteries must decline further if EVs are to win over consumers worried about range anxiety, even if those worries are unfounded. Tesla and other companies have already focused on reducing the cost of batteries, but further cost savings are necessary to extend the range of electric vehicles.

The other solution to range anxiety is to expand the number of high-voltage charging spots available to drivers. Surveys in California, home to 30 percent of electric vehicles in the United States, reveals that a majority of EV drivers are concerned about a shortage of charge stations.[32] In 2017, American drivers of ICE vehicles could refuel at any of over 121,000 gas stations.[33] By way of contrast, drivers of EVs had only 16,000 charging stations to choose from, and not all of those stations are compatible with every model of electric vehicle.[34] In reality, the maximum distance between charging stations is well within the range of EVs, but drivers have been spoiled by the great number of gasoline stations, creating an impression that there is a shortage of places to recharge EVs. Changing consumer perceptions about range anxiety will require a massive expansion in the availability of public charging stations.

BACK TO THE FUTURE

Henry Ford and Thomas Edison failed in their attempt to create an electric vehicle for the mass market. A century later, the Ford Motor Company is investing billions of dollars in EVs. In 2017 Mark Fields, CEO of Ford, announced that the company's "investments and expanding lineup reflect our view that global offerings of electrified vehicles will exceed gasoline-powered vehicles within the next 15 years."[35] Ford is also investing in high-speed electric vehicle charging stations in Europe and developing a wireless charging station in the United States.

The connection between the rise of electric vehicles and the rise of renewable energy is not self-evident. It is a process known as "convergence," and it is likely to be critical to the success of both sectors. But before addressing convergence between renewables and EVs, it is important to first understand the status of each sector as they approach what is called "parity."

FIGURE 9.1 Warren Buffett. (Source: Wikimedia Commons)

9

PARITY

We have got a big appetite for wind or solar. If someone walks in with a solar project tomorrow and it takes a billion dollars or three billion dollars, we're ready to do it. The more there is the better.
—Warren Buffett, Berkshire Hathaway
Annual Shareholder Meeting, 2017

GRID PARITY

Parity describes the point at which the price of electricity generated from renewable energy sources equals the price of electricity from fossil fuels. Often referred to as grid parity, proponents of renewables have long considered it a key objective for the industry, as its achievement would signal that renewable energy can compete with fossil fuels.

Grid parity is a powerful concept because electricity is a commodity. Consumers of electricity, including households, businesses, and industry, have shown a propensity to switch the source of power generation from fossil fuels to renewables when the price is equal to or lower than what they are currently paying to their electric utility, which is the grid price. Recall from chapter 6 the introduction of solar leasing, a financing product that allows consumers to place solar panels on their homes without making any up-front investment. Solar leasing companies have found that consumers are likely to switch from fossil fuel–sourced electricity from the utility to solar power on their roof when offered a saving on their monthly utility bill.[1] Lyndon Rive, founder of SolarCity, the largest solar leasing company in the United States, succinctly summed up the attraction of offering solar at a cost below grid parity: "If you have the choice of paying more for dirty energy, or less for clean energy, which one would you pick?"[2]

Electricity generated from renewable wind and solar has reached grid parity in some places but not others. Electricity prices range by a factor of two or three in the United States alone, and this figure is significantly higher in other countries. And the cost of electricity from wind and solar is, as discussed earlier in this book, highly dependent on location. Therefore, the concept of grid parity is only useful, in a practical sense, depending on where it is measured.

A study in 2016 found that residential solar had reached grid parity in twenty of fifty U.S. states, meaning that homeowners could purchase electricity from solar panels for equal to or less than the cost from the electrical grid (figure 9.2). In some states, like California, this is due to high fossil fuel–based electricity prices from the grid combined with abundant sunshine to power

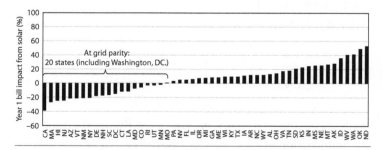

FIGURE 9.2 U.S. states at grid parity in 2016. (Adapted from Cory Honeyman, "U.S. Residential Solar Economic Outlook 2016–2020: Grid Parity, Rate Design and Net Metering Risk," GTM Research, https://www.greentechmedia.com/research/report/us-residential -solar-economic-outlook-2016-2020#gs.aTprBdE)

solar installations. At the other extreme, states like North Dakota have less sunlight, given their northern location, and very low electricity rates, making it untenable for homeowners to generate savings from solar power at current prices.

As the cost of solar and wind power continues to decline, additional regions reach grid parity, and more consumers switch from fossil fuel–based electricity to renewable energy. This drives demand for more renewables, the learning curve leads to lower production costs, and the virtuous cycle repeats itself. However, the concept of grid parity turns out to be more complex than it first seems.

THE GRID PARITY DEBATE

The simplicity of grid parity can be misleading. Electricity is a commodity, but the price that consumers pay for electricity is only one measure of the competitiveness of the source of that energy. All forms of energy production enjoy various subsidies, direct and indirect, that affect the final price paid by consumers. Renewable energy, including wind and solar power, receives government subsidies in many countries, as described in chapter 6. Similarly, electricity generated from fossil fuels is supported by government subsidies. In 2014, global subsidies for fossil fuels totaled $452 billion, including $20 billion in the United States alone.[3] The International Energy Agency estimates that subsidies to fossil fuels are almost four times the subsidies provided to renewables.[4]

Despite frequent controversy and debate among politicians, government subsidies are relatively easy to measure, and the calculation of grid parity can be adjusted to account for subsidies. In fact, most reports on grid parity make the calculation with and without subsidies. More importantly, the debate over government subsidies misses two key issues surrounding grid parity, both of which are significant and challenging to calculate: externalities and intermittency.

- *Externalities*: Burning fossil fuels to generate electricity emits gases, such as CO_2, which pollute the atmosphere and contribute to climate change. Pollution is an externality— an economic term for the side effect of an activity that is not reflected in the price of the goods or services being produced. Generating electricity with fossil fuels creates several negative externalities, which impose a burden and a cost on society but not on the polluter. In addition to climate change, burning coal emits sulfur dioxide, causing acid rain and mercury, a toxic chemical. It is in this way that the true costs of energy from fossil fuels are not captured in the price paid by consumers on their electric bill. Renewable wind and solar, on the other hand, emit no pollutants, and therefore do not create any negative externalities or costs. Unfortunately, the calculation of grid parity does not account for the cost of negative externalities created from burning fossil fuels.

- *Intermittency*: Wind and solar generate electricity only when it is windy or sunny, yet consumers of electricity expect and rely on a consistent availability of power. Fossil fuels and nuclear can generate what is referred to as baseload, which provides a fixed amount of power twenty-four hours a day. And as demand for electricity varies throughout the day, natural gas–fired plants are very effective at providing dispatchable power, increasing or decreasing available electrical capacity as needed. However, the calculation of grid parity does not consider the value fossil fuels provide in supplying both baseload and dispatchable power. Put another way, the intermittency of renewable energy must be balanced with more consistent sources of electricity, which is not captured in the calculation of grid parity.

Grid parity is an imperfect concept, one that produces debate between proponents of renewable energy, who decry the cost of negative externalities arising from the burning of fossil fuels, and proponents of fossil fuels, who point to the implicit value of providing baseload and dispatchable power. Nevertheless, grid parity is an important concept because electricity is a commodity, purchased primarily on price. When the cost of electricity from renewable energy generation reaches parity with the cost of electricity from fossil fuels, consumers switch energy sources. More importantly, the occasionally heated debate over the concept of grid parity may be missing the big picture—the cost of renewable energy continues to decline, making renewables not just equal in price to fossil fuels but increasingly less costly, taking the sector beyond grid parity.

BEYOND GRID PARITY

In many places the cost of electricity generated from wind and solar has declined to the point where it is not just competitive with fossil fuels but significantly cheaper. A combination of new technologies and increasingly efficient manufacturing is the driver of this trend. In the solar sector, utility-scale solar projects are integrating trackers—equipment that enables solar panels to adjust position to track the sun throughout the day—thereby increasing electricity generation by up to 6 percent and further reducing the LCOE.[5] Complex algorithms and machine learning support tracking in real time, adjusting panels as the angle of the sun changes throughout the year. Industry forecasts are for nearly 50 percent of ground-mounted solar projects to incorporate trackers by 2021.[6]

The wind sector has also incorporated newer technologies to improve efficiency and reduce the cost of electricity. Gearless turbines and magnets made of rare earth metals have improved the reliability of wind turbines. Innovations in blade design alone have led to an increase in energy capture of 12 percent.[7] Most importantly, turbine sizes continue to increase. The most common size for wind towers in the United States is currently 1.5 MW, with prototypes for up to 15-MW turbines under development.[8] These massive machines can be placed offshore, where wind speeds are stronger and more constant.

HOMEOWNERS AND THE ALLURE OF DISTRIBUTED ENERGY

Since electricity is a commodity, consumers are generally indifferent about its source, as price is the primary determining factor. Many consumers, however, are strongly attracted to the idea of distributed energy, which is electricity generated at the consumer's home or business. Solar is the most popular form of distributed energy, as it can be configured to the rooftops of most buildings.

The appeal of distributed energy is as much psychological as it is economic. It turns out that most homeowners are wary of their electric utilities. A study by the consulting firm Accenture revealed that fewer than one-fourth of customers trust their utility.[9] Homeowners prefer to own and control the power they use if the cost is equal to or lower than the cost of electricity from the utility. Solar power provides an ideal way for homeowners to act on their distrust of utilities.

Another psychological aspect of renewable energy is referred to as the contagion effect. Proximity to neighbors who install solar panels increases the likelihood that other homeowners will also install them.[10] As more households install renewable energy, the perception of risk changes: what was once perceived as a novel and potentially risky source of power becomes the new normal. The homebuilding industry has taken notice. Nearly every one of the top ten U.S. homebuilders includes solar panels in its offerings, either as an add-on or standard feature in new construction.[11]

When the cost of renewable energy falls below grid parity, it provides an economic incentive for homeowners to act on their psychological preference for distributed renewable energy over power provided by the local fossil fuel–based utility. The adoption rate of renewable energy by homeowners in places where it has breached grid parity is remarkable. In Australia, residential solar penetration went from near zero to 23 percent of the entire country in just eight years.[12] In some Australian towns, nearly 70 percent of households have installed solar panels.[13] Rapid implementation of solar by consumers is not unique to sunny Australia. Homeowners in California have installed solar panels on 8 percent of all residences, primarily because it is cheaper than grid power.[14]

This growing shift toward decentralized, renewable energy generation has created a class of energy consumers called "prosumers." These are homeowners or businesses that consume and produce energy generated via solar panels on their homes or places of work. Prosumers are especially attracted to renewable energy because of the control it provides over their energy production. Prosumers maintain a connection to the electrical grid so that they can draw electricity from it when they require more energy than they produce to power their home or business. Likewise, when prosumers produce more electricity than they need, they can sell the excess back to the grid for resale to other consumers.

Solar and wind power breaching grid parity has enabled homeowners to act on their desire to take control of their energy sources while also saving money. While these prosumers were the first to discover the benefits of renewables, corporations are now joining this movement as well.

CORPORATE COST SAVINGS

In 2015, Apple Inc. announced a goal to use 100 percent renewable energy worldwide, and by 2017 it had nearly achieved its objective, sourcing 96 percent of energy from renewables. Apple, the world's most valuable company by market capitalization, has given several reasons for converting to renewable energy, but the primary driver is cost savings.[15] Apple can secure very low rates from renewable wind and solar projects by purchasing electricity with long-term, fixed-rate contracts. And Apple is not alone in doing this.

Walmart, a company famously focused on reducing costs, was sourcing 25 percent of its electricity from renewables in 2015, and it announced in 2016 that it will increase its use of renewables to 50 percent of its electricity needs by 2025, with a longer-term plan in place to reach 100 percent renewable energy.[16] According to Walmart CEO Bill Simon, "It's a business decision. The renewable energy we buy meets or beats prices from the grid."[17] In addition to Apple and Walmart, over one hundred multinational companies—including Coca-Cola, Facebook, Citi, GM, H&M, J&J, Nike, and Starbucks—have announced plans to transition to 100 percent renewable energy in the coming years. Given that the private sector is responsible for half of global energy consumption, these commitments have the potential to drastically increase demand for renewable energy globally, thereby further decreasing the cost.

UTILITIES AND LOW-COST POWER
GENERATION

The growth in demand for renewable energy from homeowners and corporations receives most of the publicity in the sector, but the biggest change lies within the utility industry itself. In many countries, utilities have been encouraged by government regulation to purchase renewable energy, as described in chapter 6. But when the price of power from renewable energy reached parity with fossil fuels, utilities began to source electricity from increasingly large wind and solar projects without additional government incentives, which further reduced the LCOE of electricity from renewable sources. Wind and solar projects can be built in a matter of months, compared to the years it takes to source, permit, and develop coal and natural gas plants, making it easy for utilities to move quickly to take advantage of low prices. In a remarkably short period of time, utility-scale solar and wind projects have become the preferred source of new power generation. By 2016, solar and wind power accounted for 65 percent of new energy generation in the United States, compared with 29 percent from natural gas.[18] In most countries, utilities are now building more electricity-generating capacity from renewables than from any other source of power.[19]

In the solar sector, utility-scale projects are beating plants fired by natural gas and coal in auctions to supply electricity, generating power at shockingly low prices *without government subsidies*. In 2017, electricity from utility-scale solar projects was contracted in India for the local currency equivalent of 3.9 cents per kWh, a 40-percent drop from the previous year's record-low price, and well below the average rate charged by coal plants of 5.0 cents per kWh.[20] In 2016, Chile's power auction attracted solar companies with a global record-low bid of 2.91 cents per kWh, at which price the prevailing company will supply power to the country for twenty years.[21] Only one month later the record low was again revised down, to 2.42 cents, for a power-purchase agreement on a solar project in Abu Dhabi.[22]

Utilities are also taking advantage of low electricity rates in the wind sector. At India's first wind auction, in 2017, a wind producer won with a bid of 5.4 cents per kWh,[23] slightly less than the cost of coal-fired electricity. In the United States, wind energy prices have declined to an average of 2.4 cents per kWh, well below the average wholesale price for electricity.[24] But the most promising growth opportunity for utility-scale wind projects is offshore. In many parts of the world, the strongest and most consistent winds can be found over water, where wind developers have built several of the biggest wind farms. The first offshore wind farm was built in the waters near Denmark, and the trend has spread across Europe, with large wind farms now found in the English Channel and the North Sea.[25]

Utilities around the world are attracted to large, utility-scale wind and solar projects, which can produce significant amounts of power at extremely low cost. Even without government subsidies, these renewable energy projects generate electricity at rates that are below grid parity, outcompeting fossil fuels on the metric that matters most—price. And that is attracting investors.

CAPITAL ENTERS THE SECTOR

Electricity was described earlier in this chapter as a commodity. Capital is a commodity as well. Investors seek to place capital in ways that maximize what is referred to as risk-adjusted return. In other words, investors require a higher expected return when investing in assets that they perceive to be riskier. The price of capital, therefore, is simply the risk-adjusted return demanded by investors. And, like all other commodities, investment dollars are fungible. This means that investors quickly move their capital from more expensive to less expensive assets, just as consumers of electricity move their source of power from a high-price source to one offered at a lower price.

In new investment sectors such as renewable energy, the fungible nature of capital initially creates a problem. Investors perceive new sectors to be risky given a lack of historical performance data, and they consequently require a high return if they are to allocate capital to the sector. Furthermore, the capital markets are quite efficient, especially in developed countries like the United States, meaning that investors mostly reach the same conclusions about investment opportunities. If one experienced investor perceives a sector to be risky, most other investors will reach the same conclusion, and capital will not flow to that sector unless returns are expected to be very high.

But the reverse is also true. Once investors recognize that a sector is low risk, their return expectations decline. When investor risk–adjusted return expectations decline below the forecasted returns in the new sector, capital begins to flow to the assets in that sector. And, like all commodities, once the capital begins to flow, it tends to do so in large volumes.

This explains why investors such as Warren Buffett's Berkshire Hathaway are moving capital into the renewable energy sector. Berkshire Hathaway Energy owns wind, solar, geothermal, and hydroelectric assets. Renewable and noncarbon sources account for 39 percent of their owned and contracted generating capacity.[26] In 2016, Berkshire Hathaway's energy business contributed nearly ten percent of its overall $24 billion in profit. Buffett's comment at

the Berkshire Hathaway 2017 shareholder's annual meeting
that "we have got a big appetite for wind or solar. . . . The more
there is the better" reflects the fact that investing in renewable
energy is now attractive on a risk-adjusted basis.[27] Unsurprisingly, Buffett is not alone. In the middle of 2017, the largest
asset manager in the world, BlackRock, raised $1.5 billion from
pension funds and other institutional investors and invested
in wind and solar projects; Goldman Sachs and other leading
investment firms are doing the same.[28] The renewable energy
sector has attracted more than $230 billion in investment capi-
tal every year since 2009.[29] To put that into perspective, more
money is invested globally in renewable energy than in any
other single source of electricity generation, including nuclear,
coal, natural gas, and oil.[30]

Reaching grid parity has fundamentally changed the appeal
of wind and solar energy to homeowners, businesses, and utilities, driving demand for wind and solar energy. At the same time,
institutional investors have recognized that the risk-return profile
of wind and solar projects is attractive, and they have significantly
increased capital allocated for project development. A trifecta of
lower costs, higher demand, and attractive financing has enabled
dramatic growth in the renewable energy sector. Coincidentally,
a similar story is about to play out in a separate but not unrelated
sector—electric vehicles.

PARITY IN ELECTRIC VEHICLES

The concept of parity also applies to electric vehicles. Historically, EVs have been more expensive than vehicles using internal
combustion engines due to the high cost of batteries. But the
cost of batteries is declining rapidly, as described in chapter 8.

At what point does the cost of an EV reach price parity with a gasoline-powered automobile?

Answering this question is important because price parity is a likely inflection point in the growth of demand for electric vehicles by consumers and especially by business and industry. Vehicles are not commodities like electricity because consumers choose their cars based on a multitude of factors, but the sticker price is nonetheless a major determinant of demand. When the price of EVs declines to the point of parity with gasoline-powered vehicles, demand is forecast to surge, especially given that EVs are much cheaper to maintain and operate over the course of their working life.

INFLECTION POINT?

Vehicle prices vary by country, and no two car models are identical, but several recent studies suggest that a rapidly approaching price parity with similar classes of gasoline-powered vehicles will soon lead to an inflection point in the sale of electric vehicles. The Swiss bank UBS forecasts electric vehicle prices to be comparable with those of internal combustion engines as early as 2018 in Europe, by 2023 in China, and by 2025 in the United States.[31] Analysts at Bloomberg and at the investment bank Morgan Stanley forecast price parity in the United States and Europe by 2025.[32] Most analysts expect price parity between electric vehicles and gasoline-powered automobiles between 2020 and 2025, and there is near universal agreement that when this occurs it will create dramatically greater demand for EVs. UBS illustrates the importance of price parity to the growth in EV sales in figure 9.3, forecasting compound annual growth rates of 46 percent from 2020 onward.

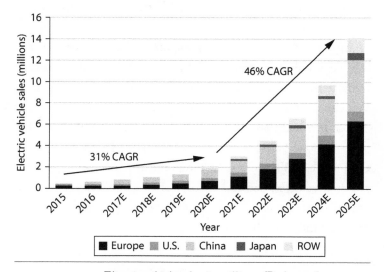

FIGURE 9.3 Electric vehicle sales, in millions. (Redrawn from "UBS Evidence Lab Electric Car Teardown," UBS Limited, www.advantagelithium.com/_resources/pdf/UBS-Article.pdf)

PARITY TOMORROW, DESIGN TODAY

In 2017, one-quarter of all new car models introduced in the United States were electric.[33] But sales of EVs in the prior year totaled less than 1 percent of new cars sold.[34] This odd discrepancy between sales and new models is a function of the long lead time between design of new vehicles and the time it takes those vehicles to reach the showroom. In the automobile industry, it can take six years to design a new vehicle, and analysts predict that six years from now EVs will be approaching price parity with traditional gasoline-powered automobiles.[35] This means that car companies need to launch new electric vehicles today if they want to remain competitive in the future. That is precisely what they are doing.

In July 2017, Volvo became the first major car company to announce that its new models will all be hybrid or electric by 2019.[36] This was quickly followed by a burst of similar announcements from leading automakers. Volkswagen, the largest automaker in the world, plans to offer eighty new electric models by 2025 and either hybrid or all-electric versions of all three hundred of its models by 2030.[37] The head of product development at the largest American automobile company declared that "General Motors believes in an all-electric future."[38]

The rise of EVs is, to a certain extent, self-fulfilling. A study by the Consumer Federation of America found that 70 percent of young adults would "consider an electric vehicle if it costs the same as a gas-powered car, has lower operating and maintenance costs, has a 200-mile range between charges, and can recharge in less than an hour."[39] By 2017, several electric models had already met these requirements. When it came on the market, the Nissan Leaf had a starting price just over $30,000.[40] The efficiency of electric motors combined with the higher cost of gasoline versus electricity make it more affordable to charge an EV than to fill a similar car with gasoline. A 2018 study from the University of Michigan found that EVs cost less than half as much in fuel costs as ICEs, averaging $485 per year versus $1,117 for gasoline-powered vehicles.[41] And an EV is cheaper to maintain because it has fewer moving parts.[42] The all-electric Chevy Bolt, Tesla S, and Tesla X all have ranges of two hundred miles or greater, and 80 percent of the battery's capacity in these cars can be recharged in thirty minutes.[43]

The study that found young adults are likely to consider purchasing an EV also found that the more consumers know about electric vehicles, the more likely they are to consider purchasing one. As automobile companies launch increasingly competitive EVs, sales will begin to accelerate, increasing consumer awareness and changing the public perception of these automobiles. Price parity between EVs and gasoline-powered vehicles will likely be the inflection point for the transition to an all-electric future.

While most research focus is on automobiles, electric vehicles' dominance over gasoline-powered vehicles might first occur in the heavy-vehicle segment of the market.

ELECTRIC TRUCKS AND BUSES

In late 2017, Tesla unveiled plans for a heavy-duty, all-electric truck with a range of 500 miles per charge. Despite the flurry of attention surrounding Tesla's announcement, the company is not the first to promote large electric vehicles, which have quietly experienced significant growth on the back of two significant advantages over diesel- or gasoline-powered vehicles: zero emissions and lower operating costs.

New Delhi, India, is replacing diesel-powered buses with zero-emission electric buses as the city grapples with debilitating smog. In November 2017, pollution levels in the city exceeded thirty times the World Health Organization's recommended maximum, resulting in a public health emergency, school closures, and grounded flights. That same month the city decided to replace five hundred public buses with all-electric models.[44] Similarly, the city of Shenzhen, China, made the decision to transition all of its fourteen thousand municipal buses to electric vehicles by the end of 2017.[45] Electric buses provide congested cities with a public transit option that reduces smog, a major benefit in cities throughout Asia suffering from poor air quality. Moreover, electric buses can save cities money.

New York City commissioned a study by Columbia University to examine the cost of replacing diesel buses with electric models.

The study found that while the initial purchase cost of electric buses would be $300,000 more per bus, the overall cost of purchasing and operating an electric bus would be $168,000 lower due to lower fuel and maintenance costs.[46] Those savings did not take into consideration lower health costs for city residents and the public health care programs many residents rely on. Diesel engines emit particulates, which are responsible for a range of respiratory and heart diseases. The same report forecasted annual savings on health costs of roughly $100 for every New York City resident if diesel buses were replaced with electric ones. Unsurprisingly, the city is now piloting its first electric buses.[47]

In the truck sector, interest in electric vehicles is primarily based on potential cost savings. Tesla announced that its all-electric truck would operate at a cost of $1.26 per mile, compared with $1.51 per mile for a diesel truck, generating material savings for fleet owners in the highly competitive trucking industry.[48] Analysis by global consulting firm McKinsey & Company shows that light-duty electric trucks operating on routes of fewer than 130 miles are already cost competitive with diesel vehicles in Europe and the United States.[49] And Mercedes-Benz is piloting an electric truck capable of hauling twenty-six tons to compete with the most powerful long-haul trucks.[50]

BUILD YOUR DREAMS

In 2008, a subsidiary of Warren Buffett's Berkshire Hathaway invested $230 million in return for a 10 percent stake in BYD, a little-known Chinese battery manufacturer.[51] BYD stands for Build Your Dreams. The vision of the company's founder, Wang Chuanfu, is to produce electric vehicles that can compete head on with the internal combustion engine. No company has positioned itself more aggressively for electric vehicle parity than BYD, and no company, not even Tesla, has come as close to reaching that goal.

Established in southern China, BYD began as a manufacturer of rechargeable batteries. The company grew rapidly, becoming China's largest producer of batteries before expanding into electric vehicle production. By 2017, BYD had seven vehicle models and was the best-selling electric car company in China, the world's largest auto market.[52] BYD also has aggressively expanded into manufacturing electric buses and light-duty electric trucks. The company is supplying the city of Shenzhen with fourteen thousand electric buses with which to replace its entire diesel fleet. As for Warren Buffett's investment in BYD, it appreciated nearly tenfold to $2.2 billion.[53]

THE RELATIONSHIP BETWEEN ELECTRIC VEHICLES AND RENEWABLE ENERGY

Grid parity is accelerating the growth of electricity generated from wind and solar, and price parity will soon accelerate the growth of electric vehicles. The symbiotic relationship between these two massive industries—electric power and transportation—is referred to as convergence. Convergence is a critically important phenomenon, providing a path for addressing the energy storage problem facing renewable energy in the twenty-first century.

FIGURE 10.1 An electric car powered by solar energy. (Photo courtesy of SunPower, https://us.sunpower.com/blog/2015/09/12/electric-cars -and-solar-power-go-better-together-green-lifestyle/)

10

CONVERGENCE

*By 2025, everybody will be able to produce and store power.
And it will be green and cost competitive.*

WIND and solar power produce low-cost electricity, but because they are intermittent the electricity generated must be stored if renewables are to replace fossil fuels entirely. Coincidentally, electric vehicles are essentially batteries on wheels, capable of drawing electricity from the electric grid, storing power, and using it when needed. Linking renewable energy with electric vehicles, a process called convergence, simultaneously solves two problems—it eliminates the intermittency problem for wind and solar while reducing the cost of electric vehicles. Convergence is a simple idea based on complex economics. Understanding the economics requires an explanation of how electricity is managed on the grid.

CALIFORNIA'S DUCK CURVE

Electricity is consumed in varying amounts throughout the day; these levels are described by a load profile—a graph of demand for electricity in a twenty-four-hour day. Electricity demand is typically lowest during the night, when most people are sleeping and businesses are closed. During the day, the load profile increases, especially in warmer regions with air conditioning. The load profile typically reaches a peak, referred to as peak load, in the late afternoon or early evening, when air temperatures reach a daily high and workers return home to switch on air conditioning, lights, appliances, and other electrical devices.

Managing the load profile is an essential task of a modern electrical grid, as consumers and businesses expect electricity to be continuously available. Fluctuating demand for electricity requires the grid operator to supply enough electricity to meet peak load without oversupplying electricity when demand falls. This challenge becomes much greater when the generation of electricity is from intermittent sources. To further complicate matters, wind energy generation is typically highest at night, when winds are strongest; these times are, conversely, when demand for electricity is at its lowest. Solar energy generation is highest at mid-day, declining in late afternoon and evening as the sun sets, when demand for electricity is usually peaking.

The challenge posed by the intermittency of wind and solar power is already evident in California, where generation of electricity by renewables has affected the entire state's load profile for electricity. Figure 10.2 is the load profile for California.[1] Demand for electricity, as shown by the line labeled "Total load," hits a low at about 5 a.m., then slowly increases throughout the day, peaking at about 7 p.m. The problem that has emerged is that generation of electricity from solar, as shown by the "Solar output" line, peaks around two o'clock in the afternoon. This distorts the net demand curve—the difference between overall demand and solar power output—to the line labeled "Load-solar-wind." This distorted load profile vaguely resembles the profile of a duck, hence it is referred to in the utility industry as the "duck curve."

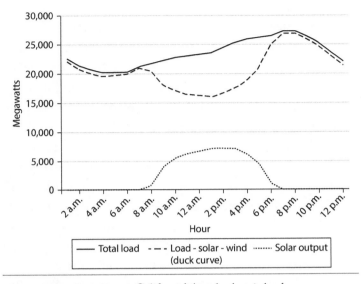

FIGURE 10.2 California's hourly electric load.
(Source: Wikimedia Commons)

The grid operator must ensure that the supply of electricity precisely meets the demand curve throughout every twenty-four-hour day, a challenge made increasingly difficult by the growing supply of electricity from wind turbines and solar panels. The solution to the duck curve is to store electricity generated when it is not needed, mostly in the middle of the day, and then use it in the early evening after the sun sets. In fact, the storage of electricity at grid scale has already been implemented in many countries.

ELECTRICITY STORAGE SOLUTIONS

More than 96 percent of global grid-scale storage capacity is in the form of pumped hydro storage, a technology developed over one hundred years ago.[2] Pumped hydro storage uses electricity to pump water from a lower elevation reservoir to a higher one. When electricity is needed, the process is reversed, and stored water from the higher reservoir is released through turbines, which generates electricity. Pumped storage hydro is a reliable, dispatchable system for producing electricity on demand, with a round-trip efficiency that converts 70–75 percent of the electricity used to pump water back to the grid when needed.[3]

Unfortunately, pumped hydro is a poor solution to the storage challenge resulting from the rapid growth in renewable power generation. Pumped hydro storage is only feasible where the geology provides for two large water reservoirs conveniently separated by significant elevation, a sequence of topographic features occurring in few locations. Even more prohibitively, pumped hydro storage is costly to construct. The capital costs of building a pumped hydro storage facility are $238 to $350 per kWh.[4] This cost is currently lower than that of most other forms of electricity storage, but costs are not forecast to decline given the extensive engineering required to build pumped hydro storage sites. Technologies like pumped hydro, with engineering and design that are unique to each location, do not benefit from a learning curve to drive down costs.

Analysts at the investment bank Lazard have reviewed the many different technologies currently available for storing electricity on the grid, including pumped hydro, compressed air, multiple battery technologies, and even flywheels. In all cases, the current cost for electrical storage is quite high. Lazard's analysts also forecast the outlook for declining costs in technologies for the next five years. Unsurprisingly, battery costs are forecast to decline rapidly, as was discussed in chapter 8. Lithium-ion batteries have the advantage of being extremely efficient, exhibiting round-trip efficiency of 92–93 percent and manufacturing costs that are declining at 11 percent annually.[5] Given these advantages, batteries, especially lithium-ion batteries, are quickly becoming the preferred solution to the storage problem. Conveniently, a very large number of lithium-ion batteries will shortly be available to store electricity.

DECAPITATING THE DUCK

Electric vehicles are mostly recharged at night, when wind turbines generate large amounts of low-cost power, and during the middle of the day, when solar panels do the same. Conveniently, these are the two times of the day when most vehicles sit unused. Even better, electric vehicles also have the capability to send electricity back to the grid after drivers return home in the evening, just as the sun is setting. In other words, electric vehicles can recharge when electricity is most available on the grid and discharge to the grid when electricity is in short supply, a process called vehicle-to-grid (V2G). In this way, electric vehicles are uniquely positioned to eliminate the challenge of balancing the load profile of the electric grid, a solution referred to as decapitating the duck.

An electric vehicle can easily be programmed to recharge when inexpensive renewable power is available and discharge to the grid when it becomes cloudy or the wind drops. By plugging electric vehicles into home outlets during the night and into electric chargers at work during the day, drivers can automatically connect their "batteries on wheels" to the electric grid. The average automobile is driven forty-eight minutes per day, leaving more than twenty-three hours per day for electric vehicles to act as storage for the grid.[6] Drivers of electric vehicles will be incentivized to do this.

Nissan Motor Company and Enel SpA, one of Europe's biggest utilities, are running a trial in Denmark, paying electric vehicle owners to use their batteries to provide electricity to the grid to smooth supply-and-demand imbalances. Payments are up to $1,530 per year to electric vehicle owners, a significant revenue stream for the car owner, requiring no additional effort or expense.[7] A similar project is underway in the UK, where it is estimated that connecting Nissan's eighteen thousand electric vehicles to the grid would produce the combined power output of a 180 MW power station. This is equivalent to the output of a small natural gas peaking plant traditionally used to balance grid electricity load.[8] And a Dutch utility is working with Tesla and BMW to design software that instructs an electric vehicle to recharge its battery when there is excess power on the grid from wind or solar generation, in return for which the electric vehicle owner is charged a much lower rate for electricity.[9]

Batteries in electric vehicles have more than enough capacity to provide power for an entire home. The typical U.S. household consumes 30 kWh of electricity daily, half the 60 kWh stored in Tesla's smallest battery pack for cars.[10] With this in mind, Tesla has enabled software in its vehicles to allow the vehicle owner to sell electricity from the battery back to the grid. The promise of V2G was one of the reasons given for Tesla's 2016 acquisition of SolarCity, the leading U.S. residential solar company. Tesla's vision is one in which consumers place solar panels on their homes, own electric vehicles, and take advantage of the convergence occurring between them.

DEGRADATION

V2G technology is still grappling with the challenge of degradation, the reduction in capacity or lifespan of lithium-ion batteries that are frequently charged and discharged. Some studies have found that battery life will diminish significantly if electric vehicles are used to store grid electricity, in certain cases by up to 75 percent.[11] Other studies have shown that intelligent software can reduce degradation. A UK-based study of fifty battery degradation experiments found that a smart grid algorithm could in fact *improve* battery life by up to 9 percent.[12] While the jury is still out on battery degradation from V2G, every study on "smart charging" shows that electric vehicles can easily reduce costs while helping to load balance the grid simply by recharging when excess power is available, ideally when renewable wind or solar are generating electricity. A study by the U.S. Department of Energy found that smart charging can reduce electric vehicle costs by 25 percent while simultaneously helping to balance the electrical grid.[13]

TECHNOLOGY OPTIMIZING ENERGY STORAGE

Several technologies are under development that may significantly improve V2G charging, further reducing the storage cost of electricity. Sophisticated software, using blockchain for distributed recordkeeping, will enable electric vehicle owners to turn their automobiles and solar panels into revenue-generating assets that optimize for price, location, and timing. There is also a significant network effect with distributed renewable energy: the intermittency of wind and solar power can be partially offset by expanding the geographic area producing electricity, connected through high-voltage, direct-current transmission lines.[14]

Looking beyond the current models of electric vehicles, the rapid development of autonomous electric vehicles, with the capability to drive to recharging points as needed, will provide additional flexibility. Autonomous vehicles can be programmed not only to recharge when surplus electricity is available from renewable sources but also to drive themselves to the point on the grid where electricity is cheapest. Conversely, autonomous vehicles can position themselves to provide backup storage at the grid location, or node, where electricity is most needed, maximizing the value of the storage to the vehicle's owner. With this opportunity in mind, General Motors based its first self-driving, autonomous vehicle on the Chevy Bolt, an all-electric vehicle.[15]

EVER CHEAPER STORAGE

Tesla's Gigafactory for the production of batteries, described in chapter 8, more than doubled the entire world production of lithium-ion batteries while driving down production costs by more than 30 percent. Tesla was first, but competitors are now racing to catch up. Daimler, producer of Mercedes-Benz automobiles, is constructing a battery Gigafactory in Germany to cut the costs of lithium-ion battery packs by 43 percent.[16] And that's just the beginning. In 2017, plans for no fewer than ten Gigafactories were announced, and global battery production is on track to reach double the production levels of 2017 by 2021.[17]

Government regulations are contributing to the growth in battery storage. In 2016, California enacted a law requiring investor-owned utilities to invest in 500 MW of battery storage to keep pace with the expectation that renewable energy will provide 50 percent of the state's power by 2030. Sufficient energy storage capabilities will allow California to maintain power supplies even during periods of low wind and sun, eliminating the duck curve.[18]

Increasing demand for electric storage solutions from utilities, in tandem with increasing demand for batteries from automobile companies, is propelling a surge in manufacturing capacity. Application of the learning curve to these manufacturing facilities, combined with intense competitive pressure, is almost certain to result in lower prices for batteries globally. And ever cheaper storage will further increase demand, accelerating convergence between the wind, solar, and electric vehicle sectors.

DEMAND FOR ELECTRICITY GOES UP

The symbiotic relationship between electric vehicles and renewable energy is already clear, as electric vehicles will provide storage to eliminate the intermittency challenge from wind and solar power. But that's only one-half of the story. The growth in electric vehicles means, of course, that there is growing demand for electricity. In the UK, a study by National Grid found that by 2030 the electrical grid would need up to 13 percent more power generation due to demand from electric vehicles.[19] That additional demand for electricity will be inexpensively provided by solar and wind. More solar and wind power means lower prices as the learning curve drives down the cost of renewable power generation, and lower prices for electricity means higher demand for electric vehicles.

HURDLES TO CONVERGENCE

Convergence between electric vehicles and renewable energy may encounter several hurdles that could slow growth in either sector. The global energy system was designed to generate electricity at large, centralized utilities and then transmit it to consumers. For transportation, the energy system was designed to power vehicles by extracting and refining fossil fuels that are distributed through a broad network of retail gas stations. Replacing all that infrastructure will present significant hurdles, including the following:

• *Raw materials*: Silicon, used in the production of solar panels, is the most abundant element in the Earth's crust and is unlikely to undergo shortages. Lithium, used in lithium-ion batteries, is also abundantly available, but mining operations will need to be scaled up to meet forecasted demand. There is enough lithium in the earth to sustain current demand for 365 years,[20] but a dramatic increase in demand could result in a future shortfall; construction of one hundred Gigafactories for battery production would reduce lithium reserves to less than a seventeen-year supply.[21] Rare earth metals are used in magnets in some advanced wind turbine generators and electric motors, raising concerns that shortages of these raw materials could impede the growth in wind power or electric vehicles. However, recent analysis points out that "despite their name, rare earths are not geologically rare" and that the fear of shortages is nothing more than a persistent myth.[22]

• *Siting*: Earlier chapters in this book described the attractive economics of wind and solar farms. Yet even projects with good economics can be slowed by a small number of adversaries, sometimes referred to as NIMBYs (not in my backyard). Wind turbines require very little land, but their massive size can lead to opposition from nearby residents complaining of obstructed views. Solar panels are less visible than tall wind turbines, but land used for solar farms cannot be used for agriculture or any other purpose. Replacing a typical 1 GW coal-fired utility requires approximately seven thousand acres of solar panels. Obtaining permits to use large areas of land for solar power can be challenging, even though renewable energy does not require any land for mining, refining, or transporting fuel, as do coal, natural gas, and oil.

• *Transmission*: Moving electricity from solar and wind farms to households and businesses can require construction of additional transmission lines. Texas has more wind power generation than any other U.S. state, partly due to construction of new transmission lines from the windy western part of the state to city residents in the south and east. An $8 billion transmission line expansion was completed in 2014, and there are plans for additional lines to meet further growth in wind energy generation.[23] Unfortunately, not all states or countries are as supportive as Texas when it comes to building new transmission lines. In Germany, delays in expansion of the transmission system is slowing growth in the wind sector despite strong government support.[24] Even China is experiencing bottlenecks in transmission. Rapid growth in wind and solar power underpin China's plans to build sixteen new, ultra-high-voltage DC transmission lines to bring power from the interior to heavily populated coastal cities.[25]

• *Charging stations*: Growth in electric vehicles may be constrained due to a shortage of infrastructure for recharging. Electric vehicles can be charged using a normal electrical outlet, but this is slow, and drivers who park on the street or prefer to charge their vehicle at work need to use independent charge stations, ideally equipped with fast, high-voltage chargers. In the United States, there are over 121,000 retail gasoline outlets[26] but only 16,000 public charging stations, of which only 2,000 are high voltage.[27] Building out a high-speed charging station network that rivals the convenience of gasoline stations is essential to the convergence of electric vehicles and renewable energy.

• *Financing*: Developing renewable energy projects requires massive amounts of capital. Analysts forecast a need for more than $7 trillion in investment capital to finance growing demand for wind and solar projects through 2040.[28] This is in addition to the funding needed to build the electric vehicle assembly lines, battery Gigafactories, electrical transmission lines, and charge stations with the aim of creating convergence with renewables. Investments in clean energy globally are averaging more than $250 billion per year, but this is only half of what is needed to meet demand in the years ahead.[29]

• *Regulations*: The utility industry is highly regulated, designed around a hundred-year-old model in which electricity is generated at one location and then transmitted to a network of homes and businesses. Today, because of renewables, electricity is generated according to an increasingly distributed model, not only at centralized locations but also on building rooftops, in farmers' fields, and on millions of homes. While technology for distributed energy distribution and storage is rapidly getting better and cheaper, regulations are not. For example, in seven U.S. states selling electricity back to the grid, called net metering, is prohibited even if a home or business has generated more electricity than can be used. Many other regulatory barriers to renewable energy and electric vehicles exist.

The challenges to convergence between renewable energy and electric vehicles are significant and are likely to slow the energy transition. Yet none of the hurdles is insurmountable. The transition to wind and solar may be drawn out simply due to inertia, the unwillingness by many people to try something new. From bureaucrats to businesspeople, there is often an aversion to change, especially when it involves new technologies. Australia is a case in point.

CONVERGENCE DOWN UNDER

Wind and solar energy have grown extremely rapidly in South Australia due to low-cost installations; these renewables generate 57 percent of the state's electricity needs.[30] Unfortunately, in early 2017, the state experienced multiple electricity shortages, which the government blamed on renewable energy sources, especially wind farms. Solutions proposed by the government were to revert to the old way of doing things: build a natural gas peaking plant, burn additional coal, or restrict growth in wind and solar.

Elon Musk, CEO of Tesla, stepped into the fray with an audacious offer to provide electricity storage, using the electric vehicle batteries produced by Tesla's Gigafactory to solve South Australia's problem in record time, with no risk involved. Musk used Twitter to announce: "Tesla will get the system installed and working 100 days from contract signature or it is free."[31]

The South Australian government accepted the offer, and Tesla installed the storage system well ahead of the due date. It was an unusual but impressive demonstration of how the convergence between renewable energy and electric vehicle batteries can provide a better storage solution than conventional approaches.

The convergence between the electric vehicle and renewable energy sectors can take many forms. In South Australia, electricity storage did not come directly from electric vehicles but from low-cost batteries manufactured by an electric vehicle manufacturing company, providing an energy storage solution to allow for increasing use of renewables. Solving the intermittency storage problem in less than one hundred days is remarkably fast in an industry that traditionally moves in years if not decades, proving that the convergence of renewables and electric vehicle technology has the potential to radically and quickly upend the traditional energy system.

FROM CONVERGENCE TO CONSEQUENCES

The convergence between electric vehicles and renewables will make both sectors more competitive. Renewable power generation provides a solution to the increasing demand for electricity to power electric vehicles. And electric vehicles provide a solution to the intermittency problem of renewable wind and solar power generation, as vehicle owners can draw electricity from the grid when it is plentiful on windy and sunny days and sell it back when there is a shortfall on windless and cloudy days. Solving intermittency removes the only real barrier facing the transition from coal, natural gas, and oil to wind and solar power. This energy transition from fossil fuels to renewable energy has already begun, is gaining momentum, and will have immense consequences—geopolitical, economic, health, and environmental—on a global scale.

FIGURE 11.1 Xi Jinping, general secretary of the Communist Party of China and president of the People's Republic of China. (Source: Wikimedia Commons)

11

CONSEQUENCES

Clear waters and green mountains are as good as mountains of gold and silver.

—Chinese president Xi Jinping, 2017

ENERGY transitions are for the most part rather dull and painfully slow. The consequences are not. The energy transition from fossil fuels to renewables will lead to dramatic changes in the global energy and utilities industries, with ramifications for investors, employees, businesses, and governments. Investment value will shift across industries, as will employment. Geopolitics will adjust to a world in which oil and other fossil fuels are of declining value. Health consequences will include longer lives and lower medical costs. And, most importantly, the renewable energy transition will offer the single best path for avoiding catastrophic climate change.

GEOPOLITICAL CONSEQUENCES

Renewable energy technologies were predominately invented in the United States and Europe, where government support led to early experimentation with alternative sources of energy. Electric vehicle technology was similarly advanced, with government subsidies, technological innovations, and entrepreneurial businesses leading the way. American electric car company Tesla is emblematic of that process. But the strongest growth in renewable energy and electric vehicles is likely to be in China and India, the world's fastest growing economies for energy and transportation.

China and India face two major challenges in the twenty-first century—local air pollution and an absence of oil reserves. Renewable energy and the electrification of transportation provide a neat solution to both problems. Local air pollution is a byproduct of using coal to generate electricity. In China and India, coal accounts for more than 70 percent of electricity generation, creating near-suffocating air quality in major cities in both countries.[1] Measures of particulates in major Chinese and Indian cities are as high as thirty times the recommended maximum of the World Health Organization, giving rise to public health emergencies.[2]

Unsurprisingly, China and India are taking the lead from Europe and the United States in the transition from fossil fuels to nonpolluting renewable energy. China has installed more wind farms and solar projects than the United States or Europe and continues to gain ground. In 2017, nearly 70 percent of new power generation in China was from renewable energy, primarily solar.[3] India has been slower to build renewables but is now catching up, with a plan for 57 percent of all electricity capacity to come from renewable energy within a decade. The five largest solar farms in the world are all located in China and India.[4]

In transportation, a shortage of oil reserves is prompting both China and India to shift from gasoline to electric vehicles. China's production of crude oil has scarcely increased in the past twenty years, while consumption has grown with a strong economy. Production and consumption of oil were almost in balance in 1994, but by 2016 the country was importing nearly two-thirds of its oil consumption.[5] India has followed a similar trajectory, with flat oil production since the 1980s and booming consumption making the country import-dependent for more than 80 percent of its oil needs.[6] The high economic cost of importing oil, and the attendant geopolitical risks, are pushing both China and India to promote the transition to electric vehicles. Wang Chuanfu, founder and chairman of BYD, China's largest maker of electric vehicles, bluntly stated the case for why the country is likely to go all electric in transportation: "The most urgent reason is China's oil safety and security. China will run out of oil soon."[7]

China is now the world's single largest market for electric vehicles, with six hundred thousand all-electric vehicles on the road and plans to increase that number to five million by 2020.[8] Government policies in China are designed to provide financial and nonfinancial incentives to consumers contemplating electric vehicles. Generous tax exemptions, combined with waivers on license plate restrictions in large cities, have created strong demand.[9] By 2018, Chinese companies accounted for nearly half of all electric vehicles sold globally.[10] The government also has plans for the construction of eight hundred thousand charge points, approximately twenty times as many charging outlets as in the United States.[11,12]

In 2017, India's government announced a target that all vehicles sold in the country be electric by 2030.[13] And the Chinese government has announced that it is considering placing a ban on gasoline and diesel-powered vehicles.[14] China and India are not alone in making plans to end the era of the internal combustion engine. France and Great Britain have announced bans on fossil fuel–powered vehicles by 2040, and Germany is considering similar restrictions.[15] But China is the world's largest automotive market by vehicle manufacturing and sales, which means that changes in the Chinese market are likely to have global consequences.[16] Analysts believe that China's support for electric vehicles is the primary reason that the world's largest car companies are racing to roll out new all-electric models. It was likely no coincidence that General Motors announced it "believes in an all-electric future" shortly after the Chinese government declared it is considering banning the sale of fossil fuel vehicles after 2030.[17]

The logic of China and India's transition from fossil fuels to renewable energy is clear: they are rapidly growing economies with poor oil reserves, extreme air pollution, and the opportunity to solve both problems while leading the global growth in the wind, solar, and electric vehicle markets. The geopolitical implications of this transition are significant. Oil-producing nations will experience lower revenues as demand for oil declines. And countries like China and India that transition from importing oil to meeting all their energy needs using domestic renewables will experience the geopolitical advantages that come with energy independence.

ECONOMIC CONSEQUENCES

As with every energy transition, there will be economic consequences as renewable energy replaces fossil fuels in the twenty-first century. Impacts are already being felt by the U.S. coal industry, which has experienced declining demand and prices, and the European utility industry, in which equity share value has eroded. Nearly half of the coal mined in the United States is from companies that have sought bankruptcy protection.[18] And European utilities have written off $100 billion in assets since 2010, a consequence of the shift from coal, gas, and nuclear to renewables.[19]

Electric utilities that own generating assets powered by fossil fuels or nuclear are in a difficult position. As discussed earlier in this book, renewable wind and solar are now economically competitive with fossil fuel–based generation on an LCOE basis over the life of the assets. Furthermore, once built, wind and solar projects are always cheaper than traditional sources of power because the marginal cost of electricity production is zero. The wind and the sun are perpetually free. Ironically, this has ominous implications for many utilities.

Since wind and solar can produce electricity with virtually no operating costs, utilities always use electricity from those sources before running their coal or natural gas facilities, as those power sources require costly fuel as an input. The more solar or wind power built, the less the other sources of electricity are used, making those sources of power even more expensive on a LCOE basis. To make matters worse, utilities often raise consumer prices to compensate for the higher costs of maintaining their legacy coal, natural gas, and nuclear assets, but the higher prices in turn encourage consumers to install more solar panels on homes and businesses, thus reducing demand for electricity from the utility.

This creates what is referred to as a utility "death spiral." The more that utilities raise prices, the more customers generate their own electricity. As a result, utilities must again raise prices on the remaining customers, encouraging even more households and businesses to generate their own distributed energy. As the Oxford Institute for Energy Studies puts it, "The utility business model is broken."[20] Germany's two largest utilities, RWE and E.ON, broke themselves into two parts in order to survive, each company splitting its businesses into entities for renewable energy, on the one hand, and for legacy fossil fuel and nuclear assets, on the other.[21]

The oil sector has yet to feel the impact of electric vehicles, as these account for less than 1 percent of vehicles on the road. But the implications of a transition to electric vehicles, and a decline in demand for gasoline and diesel fuel, are a serious concern for the sector. Saudi Arabia, the world's biggest oil producer, is planning a share listing and partial sale of the state oil company, Saudi Aramco, which some analysts have interpreted as a signal that demand for oil is peaking. The *Financial Times* called oil "an industry with its best days behind it."[22] And advisors to the Norwegian Government Pension Fund, the world's largest fund, which holds shares in Exxon Mobil, Royal Dutch Shell, Total, Chevron, and Statoil, has recommended divestment of all stocks in fossil fuel companies as a partial hedge, given that the fund's primary revenues are from oil extraction. In Norway, 40 percent of new cars sold run on electricity, not gasoline.[23]

Employment Gains

Many of the economic consequences of the transition to renewable energy will be positive, especially for employment. The construction of wind and solar projects creates jobs. In the United States, more than twice as many workers are employed in the solar sector than in coal, and employment in the renewable energy sector is growing at 6 percent annually, far outstripping the fossil fuel industry.[24] Solar and wind have created jobs twelve times faster than the rest of the economy.[25] These are impressive employment numbers, although they account for only a fraction of the employment growth in China.

Renewable energy in China currently provides 3.5 million jobs, by far the most of any country. And as China cuts back on coal production, the government aims to increase employment in the renewables sector with an additional thirteen million jobs by 2020.[26] The International Renewable Energy Agency projects the addition of twenty-six million renewable energy jobs globally by 2050.[27]

From an economics and employment perspective, renewable energy is inherently attractive because it is local. Most of the investment is in the development and construction of projects, and any operating costs (for example, payments to landowners) accrue to the community. Although solar panels and wind turbines are sometimes imported, the bulk of expenditures on labor, land, and interconnection are always made where the project is built. Michigan governor Jennifer Granholm succinctly captured the economic appeal of renewables: "Instead of spending nearly $2 billion a year importing coal or natural gas from other states we'll be spending our energy dollars on Michigan wind turbines, Michigan solar panels, Michigan energy-efficiency devices, all designed, manufactured, and installed by . . . Michigan workers."[28]

HEALTH CONSEQUENCES

The negative externalities from burning fossil fuels take a toll on air and water systems and on human bodies. In the case of coal, harmful effects result from its entire life cycle: from exploration and extraction to processing, transporting, and, of course, combustion. In the United States, one hundred thousand miners have lost their lives in mining accidents since 1900, and more than two hundred thousand have died from black lung disease.[29] Harvard Medical School's Center for Health and the Global

Environment calculated that the economic burden of public health impacts from coal in Appalachian communities alone was $74.6 billion in 2008.[30] Across the United States, that same study estimates that the "life cycle effects of coal and the waste stream generated are costing the U.S. public a third to over one-half of a trillion dollars annually."[31]

The primary reason that the health consequences of burning coal for power are so drastic is particle pollution. Particle pollution occurs when microscopic amounts of matter such as soot enter the atmosphere. Because these particles are small enough to embed themselves deep into lungs, their impact on human health is well documented. A World Health Organization report concluded that exposure to air pollution is "associated with increases in genetic damage, including cytogenetic abnormalities, mutations in both somatic and germ cells, and altered gene expression, which have been linked to increased cancer risk in humans."[32] Coal-fired power plants are the largest source of air pollution, accounting for 40 percent of all hazardous emissions.[33] Yet there is cause for optimism: a study of U.S. cities found that between 2000 and 2007, decreases in particle pollution lengthened the average person's lifespan by about two months.[34]

Globally, air pollution resulted in more than 5.5 million premature deaths in 2013, making it the fourth-leading risk factor for early death.[35] In that same year, air pollution cost the global economy $225 billion, according to a report from the World Bank and the Institute for Health Metrics and Evaluation.[36] But again, there is evidence that reduced burning of coal can improve human health. Following a ban on coal in Dublin in 1990, air quality improved and respiratory mortality decreased by 17 percent.[37] The energy transition from fossil fuels to renewables will dramatically reduce air pollution, especially from particulates, leading to materially longer, healthier lives for millions of people.

Climate Consequences

In 1958, an American scientist named Charles Keeling placed a sensor he had developed near the summit of Mauna Loa in Hawaii to measure the amount of CO_2 in the atmosphere. The data that Keeling collected provided the first clear evidence that burning fossil fuels for the generation of energy was steadily accumulating CO_2 in the atmosphere. CO_2 is a greenhouse gas, meaning that it keeps the earth warm by reflecting thermal radiation, or heat, back to the planet. This greenhouse effect is a natural process that is essential to maintaining a temperature hospitable to life on earth. What Keeling discovered was that the greenhouse effect was getting stronger.

Atmospheric concentrations of CO_2 were very stable from the time that humans emerged from the Stone Age, some ten thousand years ago, until the onset of the Industrial Revolution. Watt's steam engine, described in chapter 2, and subsequent

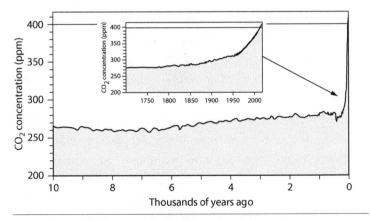

FIGURE 11.2 The Keeling Curve. (Adapted from Scripps Institution of Oceanography, "The Keeling Curve," https://scripps.ucsd .edu/programs/keelingcurve/)

advances in machinery and transportation, changed that. Steam engines were used to propel railroads, power machinery to produce goods, and generate electricity to light homes and businesses. Development of the internal combustion engine allowed for the growth of the automobile, followed by the jet engine and the age of air transport. All of this was done by burning fossil fuels—coal, oil, and natural gas—releasing vast amounts of CO_2 into the atmosphere. The Keeling Curve vividly illustrates the growth in atmospheric CO_2 concentrations since the onset of the Industrial Revolution.

The growth in atmospheric CO_2 has warmed the planet due to the greenhouse effect, and will continue to warm it as emissions rise. Every year, billions of tons of CO_2 is emitted into the atmosphere, where it remains for a century or more. As of 2018, atmospheric concentrations of CO_2 are well over 400 ppm (parts per million) and are increasing at approximately 2 ppm every year.[38] Scientists have concluded that atmospheric concentrations must be stabilized at no higher than 450 ppm to avoid "very high risk of severe, widespread and irreversible impacts globally."[39] If greenhouse gas emissions maintain their current trajectory, the planet is very likely to warm by 4 degrees Celsius by the end of the century,[40] resulting in deadly heat waves, rising oceans, and intensifying extreme weather events. Clearly, a reduction in the growth of CO_2 emissions is desperately needed. The energy transition from fossil fuels to renewable wind and solar, called "energy decarbonization," provides a path for accomplishing this goal.

Approximately two-thirds of greenhouse gas emissions are from the production and use of energy.[41] Any plan to avoid catastrophic climate change must dramatically reduce emissions from the burning of fossil fuels, which currently account for 84 percent of global energy generation.[42] Analysts from the International Renewable Energy Agency have forecast that the energy transition can take renewables from 16 percent of global energy generation up to 65 percent by 2050 by utilizing electric vehicles to provide one-half of the storage needed to manage the grid.[43] Most importantly, this same analysis finds that the energy transition to renewables would significantly contribute to the goal of keeping global temperature rise to a manageable 2 degrees Celsius.[44]

Energy decarbonization is the most significant consequence of the transition from fossil fuels to renewables. It provides a path for addressing climate change with proven technology—wind and solar—that is economic and scalable. And it can be implemented globally. But that does not mean that everyone will benefit. While the planet will surely be better off, the energy transition will not benefit everyone.

WINNERS AND LOSERS

Energy transitions go through a boom and bust cycle, creating winners and losers. Curiously, the rise and fall of the whaling industry in the eighteenth century provides interesting parallels with the current energy transition.

History of the Whaling Industry: A Cautionary Tale

The killing of whales to harvest their oil created a boom-and-bust energy transition in the United States and Europe. First discovered to be a source of lighting in the 1700s, by the mid-1800s whale oil was the lighting product of choice. Whale oil burned brightly, replacing lard oil and other lighting products that were smelly and provided less light. Great fortunes were made in whaling, and coastal towns in the northeastern United States boomed. The U.S. whaling fleet was the world's biggest, and whaling became the fifth largest industry in the country, employing tens of thousands of sailors.[45] New Bedford, Massachusetts, at the center of the American whaling industry, became the wealthiest city per capita in the United States in 1854.[46]

Then in 1859 oil was discovered in Titusville, Pennsylvania. Consumers found that petroleum was a better product than whale oil for lighting and industrial applications, precipitating the decline of the American whaling industry. As with all other energy transitions, the economic advantages of a better, cheaper product eventually overwhelmed the incumbent industry. This took time. American petroleum production was a mere two thousand barrels in 1859 versus a half million barrels of whale oil. Forty years later, with the discovery of the oil gusher at Spindletop, Americans were producing two thousand barrels of petroleum every seventeen minutes, and the demand for whale oil never recovered.[47] Investment capital moved to the petroleum industry, as did workers, and the infrastructure of whaling decayed. Whalers hung on for another sixty-eight years, making the last whaling voyage out of New Bedford in 1927.[48]

The fossil fuel sector in the twenty-first century is starting to look a lot like the whaling industry of two hundred years ago. A review of a book on whaling by the *New York Times* compared the two industries:

> Like oil, particularly in its early days, whaling spawned dazzling fortunes, depending on the brute labor of tens of thousands of men doing dirty, sweaty, dangerous work. Like oil, it began with the prizes closest to home and then found itself exploring every corner of the globe. And like oil, whaling at its peak seemed impregnable, its product so far superior to its trifling rivals, like smelly lard oil or volatile camphene, that whaling interests mocked their competitors.[49]

Twenty-First-Century Whalers

The American coal industry is following a trajectory not unlike that of the whaling industry. Employment in the coal sector has declined from a peak of more than eight hundred thousand in 1923 to fifty thousand today.[50] Some of that decline is due to automation, but coal is no longer the dominant source of electrical power generation in the country, as utilities transition to natural gas and renewables, both of which are cheaper and more competitive. In the United States, nearly half of all coal companies have gone bankrupt since 2012.[51]

It is premature to conclude that the oil industry will follow the decline in coal, but industry insiders expect major changes ahead. Ben van Beurden, CEO of Royal Dutch Shell, the world's second largest oil and gas company, believes that "the energy transition is unstoppable."[52] BP, one of the world's largest oil companies, publishes an annual outlook for the global energy market. The outlook for 2018 forecasts dramatic growth in renewables and concludes that "the mix of fuels used in power generation is set to shift materially."[53]

The shift in power generation is almost certain to upend the oil and gas sector, as is already happening in the utility sector. A report from the Institute for Energy Economics summed up the situation faced by utilities globally:

> While it may take decades yet for renewables to become the dominant form of generation globally, their presence today is permanent and their advance inevitable. Electricity utilities still considering how and when to embrace the global shift toward renewables would do well to accelerate their transition if they are to avoid the financial damage typically incurred in stranded-asset write-downs of late movers.[54]

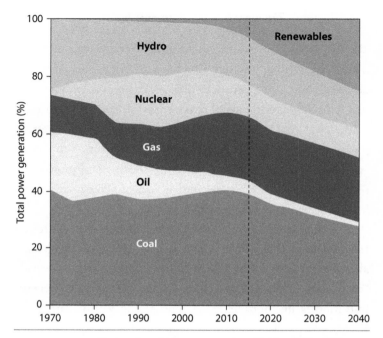

FIGURE 11.3 Share of total power generation. (Adapted from BP, *BP Energy Outlook, 2018 Edition*, https://www.bp.com/content /dam/bp/en/corporate/pdf/energy-economics/energy-outlook /bp-energy-outlook-2018.pdf)

In response to the renewable energy transition, some utilities (e.g., Enel in Italy and NextEra in the United States) are adapting to and dominating new technologies. Other companies (e.g., NRG in the United States and Tepco in Japan) are retreating into their traditional businesses, and companies on the losing end, such as RWE in Germany, have been forced into bankruptcy or reorganization.[55] As in past energy transitions, there will be business winners and losers, with vast sums of money to be gained or lost in the years ahead.

NATIONAL WINNERS AND LOSERS

Countries likely to gain from the energy transition to renewables include China and India, as discussed earlier in this chapter, and the poorest developing countries. By switching from expensive kerosene to renewable energy sources, consumers in the developing world, where 1.2 billion people lack electricity, can spend less of their income on fuel to power their homes and businesses. Developing countries also benefit from the renewable energy transition, as there is no incumbent fossil fuel infrastructure to replace. Building greenfield projects is cheaper than having to replace existing power-generation assets. Leapfrogging the use of fossil fuels and the centralized generation of electricity with decentralized renewables is a similar trajectory to the extraordinarily successful mobile phone industry, which leapfrogged landlines in developing countries.

Countries likely to lose from the energy transition are the oil-exporting nations, especially those dependent primarily on oil exports. Saudi Arabia, which, to the government's credit, is trying to diversify its economy, is aware of the risk to fossil fuels. The former oil minister Ali al-Naimi announced at a conference that "in Saudi Arabia, we recognize that eventually, one of these days, we are not going to need fossil fuels. I don't know when, in 2040, 2050 or thereafter."[56] The government is aggressively investing in renewables, especially solar given the region's abundant sunshine, with a goal to generate 10 percent of its power from renewable sources by 2023.[57]

Countries that may yet win or lose include the United States and much of Europe. In the United States, technology leadership, availability of capital, and entrepreneurship provide an ideal setting for a rapid transition to renewables. Unfortunately, this is offset by the challenges facing the construction of new infrastructure, hampered by regulations and NIMBYism, as discussed earlier in the book. The United States has also experienced unstable government policy, primarily at the federal level, where partisanship has squandered the country's leadership potential. Support for renewables has been inconsistent, while policies to encourage the extraction of fossil fuels, especially coal, are ignoring economic fundamentals. It remains to be seen if the U.S. government will support or hinder the growth of several of the world's leading companies in the renewable energy and electric vehicle sectors.

In Europe, several countries (especially Germany, but also the UK and Spain) were early leaders in the energy transition. However, the uncertainty created by Brexit and the lack of political leadership within the EU threaten Europe's standing. In summary, the United States and many European countries are well positioned to win from the transition to renewables, but near-sighted political mistakes might easily hand that opportunity to other countries.

As with all energy transitions, some countries will lead, and others will follow. Some countries will be winners, while others will be losers. The overall winner, of course, will be the planet. Potentially avoiding catastrophic climate change is by far the most important benefit of the transition to renewables. Whether or not the energy transition can achieve that goal is simply a question of time.

FIGURE 12.1 Flooding in New Orleans after Hurricane Katrina.
(Source: Wikimedia Commons)

12

NO TIME TO LOSE

The eventual response to doubling preindustrial atmospheric
CO_2 likely would be a nearly ice-free planet.
— James Hansen, Columbia University

THE key question is not *whether* the energy transition from fossil fuels to renewables is coming, but *when*. The transition is, at this point, inevitable, as the increasingly competitive economics of renewable wind and solar overtake fossil fuels. Appendix B provides a summary of the changes that have occurred, and the likely changes to come. As in the past, better economics are the primary driver of the energy transition. But this energy transition differs in one critical way from prior ones—there is no time to lose.

The previous chapter briefly described the positive climate consequences of transitioning from fossil fuels to renewables, with the potential to limit global warming to 2 degrees Celsius. This assumes, however, that the transition occurs rapidly. Understanding why time is of the essence requires a brief explanation of flows and stocks of greenhouse gases.

GREENHOUSE GAS FLOWS

Flows represent the amount of greenhouse gases emitted over a period of time, usually measured annually. This is the amount

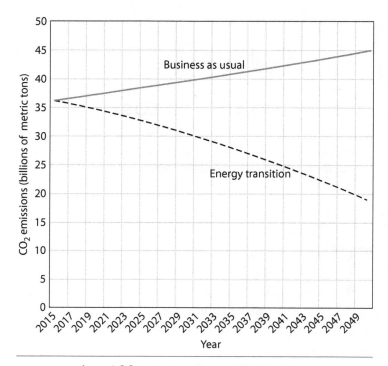

FIGURE 12.2 Annual CO_2 emissions. (Source: IRENA ReMap; author's model. See International Renewable Energy Agency, "Renewable Energy: A Key Climate Solution," November 2017, http://www.irena.org/-/media/Files/IRENA/Agency /Publication/2017/Nov/IRENA_A_key_climate_solution_2017.pdf?la=en&hash =A9561C1518629886361D12EFA11A051E004C5C98)

of CO_2 added to the atmosphere ever year. For example, global greenhouse gas emission flows in 2017 were thirty-seven billion tons of CO_2.[1] The energy transition from fossil fuels to renewables is forecast to steadily reduce CO_2 emissions, or flows, as illustrated in figure 12.2.

The International Renewable Energy Agency projects that annual emissions of CO_2, the primary cause of climate change, will decline by 58 percent by 2050 due to the energy transition to renewables and the growth in electric vehicles. This is a reason for optimism in the fight against climate change. But there is a catch. CO_2 remains in the atmosphere for approximately one hundred years, which means that the current *stock* of CO_2 will continue to warm the planet for another century, even if no additional CO_2 is emitted.

GREENHOUSE GAS STOCK

Stock represents the total amount of CO_2 accumulated in the atmosphere, commonly referred to as the concentration of CO_2. To calculate the stock, instruments are used to measure the concentration of CO_2 in the atmosphere in parts per million (ppm). This measurement, described in the previous chapter, forms the Keeling Curve. In 2018, the concentration of CO_2 in the atmosphere reached 411 ppm.[2] This measurement is critical because the greenhouse effect that warms the planet is influenced by the stock of CO_2 in the atmosphere, not by annual flows.

To avoid catastrophic climate change, it is essential that the stock or concentration of CO_2 in the atmosphere remains below 450 ppm. Scientists have determined that a concentration of CO_2 of 450 ppm is likely to warm the planet by 2 degrees Celsius, which is considered the maximum warming possible without serious risk of disruption to global economies and human welfare. Additionally, atmospheric concentrations above 450 ppm create uncertainties in feedback loops, where warming creates changes to the planet that accelerates further warming. For example, a thawing of the arctic tundra could release massive volumes of currently frozen greenhouse gases, leading to rapidly accelerating climate change and potentially catastrophic outcomes. The scientific consensus from over one thousand scientists in eighty countries is unambiguous in predicting that continued emission of greenhouse gases increases the likelihood of "severe, pervasive and irreversible impacts"[3] and in recommending that the concentration of CO_2 in the atmosphere be kept as close as possible to a maximum concentration of 450 ppm.

NO TIME TO LOSE

The challenge in staying below this maximum level is that the current concentration of CO_2 is already well above 400 ppm, and annual emissions or flows increase concentrations by approximately 2 ppm every year. CO_2 remains in the atmosphere for over a century. At current rates (labeled "Business as usual" in figures 12.2 and 12.3), CO_2 concentrations will broach 450 ppm in a little over twenty years and surpass 480 ppm by the year 2050, reaching well above the maximum threshold advised by scientists.

The renewable energy transition described in this book results in a significant decline in annual CO_2 emissions, from thirty-six billion tons today to nineteen billion tons in 2050. Compared to the business-as-usual scenario, the energy-transition scenario results in a 58 percent reduction in emissions. But even with an energy transition from fossil fuels to renewables, the atmospheric concentration of emissions will continue to climb for many years to come, partly due to the time it takes for the transition to take place and partly due to other sources of greenhouse gas emissions (agricultural fertilizers, for example).

It is essential, therefore, that the energy transition occur as quickly as possible. A delay in reducing emissions by just one decade will add another 360 billion tons of CO_2 to the stock in the atmosphere, increasing atmospheric concentrations of CO_2 by approximately 20 ppm and further accelerating the

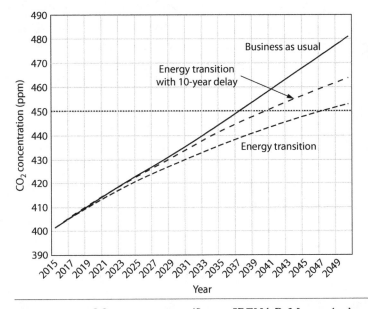

FIGURE 12.3 CO_2 concentrations. (Source: IRENA ReMap; author's model. See International Renewable Energy Agency, "Renewable Energy: A Key Climate Solution," November 2017, http://www.irena.org/-/media /Files/IRENA/Agency/Publication/2017/Nov/IRENA_A_key_climate _solution_2017.pdf?la=en&hash=A9561C1518629886361D12EFA11 A051E004C5C98)

greenhouse effect. Figure 12.3 demonstrates how a delay of only ten years takes CO_2 concentrations well above 450 ppm by 2050.

Even if the energy transition is delayed and greenhouse gas concentrations surpass 450 ppm, it would *theoretically* be possible to remove CO_2 from the atmosphere using a process called carbon sequestration. This can be accomplished by planting trees and other biomass or using experimental geoengineering techniques, such as seeding the oceans with iron. However, planting trees will require vast amounts of land that is currently used for growing food crops, and geoengineering has never been accomplished at scale. The economic cost of sequestration is unknown, and there may be severe side effects from tinkering with the planet's atmosphere.

The energy transition to renewable wind and solar provides a window of opportunity to safely limit the stock or concentration of CO_2 in the atmosphere close to 450 ppm, using proven technologies and at little or no cost, but it is very narrow and closing rapidly. A delay in the transition by just a decade closes that window, possibly forever.

TIMING THE TRANSITION

The energy transition from fossil fuels to renewable wind and solar energy is nearly certain to occur. What remains uncertain is the speed of the transition, which is key to avoiding the worst impacts of climate change. Unfortunately, predicting the timing of an energy transition is difficult. The transition from horse-drawn carriages to the automobile took a surprisingly long time—Carl Benz invented the "horseless carriage" in 1885, yet the number of cars did not surpass the number of horses in New York City until 1912, nearly three decades later.[4] In many other cities and countries it took far longer.

Will the global transition to renewable energy occur quickly enough to avoid catastrophic climate change? That answer depends, more than anything, on the price of batteries. As explained earlier in this book, batteries are the key to providing the storage necessary for transitioning from fossil fuels to renewable solar and wind power. The cost of manufacturing a lithium-ion battery declined from $750 per kWh in 2010 to just $145 per kWh in 2017,[5] but that is still too expensive—electric vehicles cannot yet compete directly with gasoline-powered automobiles, and electricity from renewable wind and solar cannot yet be stored at a cost that can compete with natural gas and coal.

At what price point must batteries be produced to ensure the energy transition from fossil fuels to renewables? The answer is less than $100 per kWh, ideally less than $75 per kWh. At that price, renewable energy with storage can compete with fossil fuels in the generation of electricity, and electric vehicles can become less expensive than conventional vehicles.[6] When will that happen? General Motors announced a goal for battery costs of $100 per kWh by 2022.[7] And Bloomberg New Energy Finance forecasts battery prices of $73 per kWh by 2030 using a learning rate of 19 percent.[8] Based on current trends, these price forecasts are likely to be achieved. This is extremely good news for the transition to renewables, and for the fight against climate change. But this fight can no longer be delayed.

THE COST OF DELAY

Higher concentrations of greenhouse gases will warm the planet, which will accelerate the melting of ice sheets in Greenland and Antarctica. Columbia University scientist James Hansen succinctly summarized the consequences of a doubling of atmospheric CO_2 concentrations when he said it would create "a nearly ice-free planet." The loss of ice has severe ramifications for humanity.

Greenland is covered in an ice sheet nearly two miles thick, which as it melts runs into the oceans and raises global sea level. The melting of the entire ice sheet in Greenland would raise sea levels by approximately twenty-four feet.[9] New York City, and about 25 percent of American homes, would be at least partially underwater.[10] But the largest risk is from Antarctica, where the ice sheet covering that continent contains enough water to raise global sea levels by over two hundred feet.[11] Once the ice sheets have melted into the oceans there is no way to reverse the rise in sea level. The most recent scientific report on global climate change concluded that these changes are irreversible once they occur.[12]

A delay in reducing greenhouse gas emissions will allow for a greater accumulation of CO_2 in the atmosphere, accelerating warming and increasing the risk of a catastrophic rise in sea level. The longer the delay in reducing CO_2 emissions, the greater the cost, as every additional ton of emissions adds to the stock already in the atmosphere and remains there for a century or more. Economists from the World Bank project the cost of flood damage to coastal cities reaching $1 trillion per year by 2050 due to rising seas.[13] Professor Gary Yohe, of Wesleyan University, describes the cost of delay more simply: "The longer you wait, the more expensive it gets."[14]

AN INEVITABLE CHANGE

In many ways climate change is a simple problem. The cause of the problem, emissions of greenhouse gases into the atmosphere, and the primary source of the problem, the burning of fossil fuels to generate energy, are well understood. Furthermore, a solution to the problem is already available: the use of renewable solar and wind energy, which are increasingly cost competitive.

What makes climate change a particularly vexing problem is that the technology already exists to transition the global economy away from fossil fuels at a cost that is declining rapidly and may already be close to zero. This is in stark contrast to the cost of delay, which is already high and continues to rise. What is lacking is the political will to ensure that the energy transition moves as quickly as possible. Governments must enact regulations that accelerate, rather than slow, the transition. That takes foresight and the willingness to promote investment in infrastructure. Most importantly, it requires governments to support a change from a world powered by fossil fuels to one powered by renewables.

Unfortunately, governments show little willingness for change; this is because not everyone benefits from change. The energy transition will create winners and losers. But a delayed energy transition will create many more losers from climate change. Inevitably, the better economics of renewables, powered by the sun and wind, will replace costly fossil fuels. Some people will wait for that to happen. Others will make it happen. All of us need it to happen, now more than ever.

APPENDIX A

Levelized Cost of Electricity (LCOE)

LCOE FORMULA

$$\text{LCOE} = \frac{\text{sum of costs over lifetime}}{\text{sum of electrical energy produced over lifetime}} = \frac{\sum_{t=1}^{n} \frac{I_t + M_t + F_t}{(1+r)^t}}{\sum_{t=1}^{n} \frac{E_t}{(1+r)^t}}$$

I_t: investment expenditures less government subsidies in the year t

M_t: operations and maintenance expenditures, including pollution abatement, in the year t

F_t: fuel expenditures in the year t

E_t: electrical energy generated in the year t

r: *discount rate* for capital investment (WACC or Weighted Average Cost of Capital)

n: *expected lifetime* of power station

EXAMPLE OF LCOE CALCULATION

Assume construction of a 100 MW wind farm with a capacity factor of 29 percent, a capital cost of $100 million, and annual operating and maintenance costs of $500,000. Assume the wind farm will operate for twenty-five years, with no residual value.

Weighted Average Cost of Capital (WACC): assume the capital cost (V) of the wind farm is financed 30 percent with equity (E) at 12 percent (Re), and 70 percent with debt (D) at 6 percent (Rd). The tax rate (T) is 30 percent.

$$\text{WACC} = \left(\frac{E}{V} \times \text{Re}\right) + \left(\left(\frac{D}{V} \times \text{Rd}\right) \times (1-T)\right)$$

$$\text{WACC} = \left(\frac{\$30}{\$100} \times 12\%\right) + \left(\left(\frac{\$70}{\$100} \times 6\%\right) \times (1-30\%)\right)$$

$$= 6.54\%$$

$$\text{LCOE} = \frac{(\$100 \text{ m in year } 0 + \$500{,}000 \text{ per year}) \text{ discounted @ } 6.54\% \text{ per annum for 25 years}}{(100 \text{ MW} \times 29\% \text{ capacity factor} \times 365 \text{ days} \times 24 \text{ hours}) \text{ per annum for 25 years}}$$

$$= \$34.36 \text{ per MWh}$$

APPENDIX B

The Transition to Renewable Energy

T HE transition to renewable energy is, at this point, inevitable, as the increasingly competitive economics of renewable wind and solar overtake fossil fuels. The following illustration provides a summary of the changes that have occurred, and the likely changes to come.

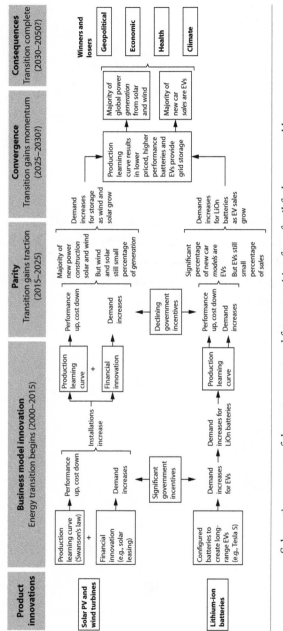

Schematic summary of the past, present, and future transition from fossil fuels to renewable energy.

GLOSSARY

BALANCE OF SYSTEMS: The components of a solar PV project other than the solar panels; generally includes a racking or mounting structure, cables, inverter, and controller and connection to the electrical grid.

BEHIND-THE-METER: A term applied to electricity used where it is generated, usually on a home or building; electricity generated behind-the-meter does not appear on the home or building's utility bill.

CAPACITY FACTOR: A measure of the electricity a power project produces compared to how much it could potentially produce if it operated at capacity; it is calculated by dividing the electricity generated by the potential generation of the project.

COMBINED-CYCLE GAS TURBINE: A power plant technology that uses both a gas and a steam turbine to generate electricity; heat that is released from the gas turbine is converted to steam to generate additional electricity.

COMMODITY: A good or service that is interchangeable with, and thus indistinguishable from, other products of the same type; electricity is one such example. Beyond price, consumers are generally indifferent about the source of a commodity.

CONCENTRATED SOLAR POWER: A type of solar plant that utilizes mirrors or other reflecting lens to concentrate the sun's rays, where they are converted to heat used to drive a steam-powered turbine to generate electricity.

CONVERSION EFFICIENCY: The ratio between the useful output of an energy conversion device and the input.

COST OF CAPITAL: The required return (measured as a percent) on capital invested in a project, such as a power plant.

DISCOUNTING: A process to calculate the present value of a stream of future cash flows by accounting for the time value of money.

DISPATCHABLE ENERGY: Energy that can be generated when required and is thus always available.

DISTRIBUTED GENERATION: Energy generated near to where it will be consumed; solar power constitutes a form of distributed generation when solar panels are placed on homes and buildings.

ELECTRICITY: A type of energy generated when electrons flow from one place to another; it is measured in watts per hour and expressed in kilowatt hours (kWh), megawatt hours (MWh), and gigawatt hours (GWh).

ENERGY DENSITY: The amount of energy that a substance or system can store per unit volume or mass.

EXTERNALITIES: A side effect or consequence of a commercial activity that has costs to individuals or groups other than those performing or benefiting from the activity—for example, air pollution from the burning of fossil fuels. Most externalities are negative.

GIGAWATT (GW): A unit of power equal to one billion watts; useful for measuring total global output of an energy source (e.g., the total installed capacity of wind turbines approximately doubled from 238 GW to 487 GW between 2011 and 2016).

GRID PARITY: The state achieved when the cost of generating electricity from renewable energy is no higher than the cost of electricity generated from fossil fuels.

INTERMITTENT ENERGY: Energy that is not continuously available for use as electricity; wind and solar energy are intermittent because they are only generated when the wind is blowing or the sun is shining.

INTERNAL COMBUSTION ENGINE (ICE): An engine that burns gasoline or diesel fuel to create heat inside one or more closed chambers or cylinders.

KILOWATT (KW): A unit of power equal to one thousand watts.

LEARNING CURVE: An economic concept referring to the decline in manufacturing costs of a product as the production volume increases; often measured as the percentage decline in cost for every doubling in total volume produced.

LEVELIZED COST OF ELECTRICITY (LCOE): A standard metric to compare the cost of electricity from different sources of generation; often used interchangeably with levelized cost of energy.

LOAD PROFILE: The amount of electricity consumed by a population throughout a given period of time; normally calculated over a twenty-four-hour period.

MEGAWATT (MW): A unit of power equal to one million watts; useful as a measure of the output of a power plant.

NET METERING: A billing mechanism that pays homeowners with solar systems a financial credit from the grid operator or utility for sending excess electricity to the grid.

PAY-AS-YOU-GO (PAYG): A solar financing model that allows customers to pay for their solar systems over time; used in developing countries to provide very small solar PV systems to low-income households.

PEAK LOAD: The maximum point of the load profile; the maximum amount of electricity consumed by a population over a given period of time; generally peak load refers to maximum electricity demand during a twenty-four-hour period; it may also refer to the same measure over an entire year.

POWER: A measure of the capacity to produce energy; electrical power is expressed in kilowatts (kW), megawatts (MW), and gigawatts (GW).

POWER PURCHASE AGREEMENT: A contract between the generator and purchaser of electricity, usually at a fixed price for a contractual period of twenty years or more.

RANGE ANXIETY: A driver's fear that the battery in an electric car will run out of power before reaching the destination or a place to recharge.

RENEWABLE PORTFOLIO STANDARD: A market mechanism by which governments set a minimum percentage of electricity required to be generated from renewable sources and establish a penalty for noncompliance; generally applies to utilities operating in a state.

SOFT COSTS: The costs of completing a renewable energy project other than equipment and construction; generally includes project design, permitting, legal fees, regulatory fees, labor, and financing costs.

SOLAR LEASING: A solar financing model that allows homeowners and building owners to utilize and pay for the electricity generated by the solar project mounted on the home or building roof, without having to pay for installation or maintenance of the project.

SWANSON'S LAW: The observation that the cost of producing solar PV panels declines approximately 20 percent for each doubling in cumulative production; the learning curve for solar PV panel production.

WATT (W): A unit of power, useful as a measure for household appliances (e.g., a 100-watt light bulb requires 100 watts of power to illuminate); see also *kilowatt*, *megawatt*, and *gigawatt*.

WEIGHTED AVERAGE COST OF CAPITAL (WACC): The cost of capital taking into account the relative percentage of debt and equity used to finance a project; usually also includes the tax benefit of using debt.

NOTES

PREFACE: SETTING THE RECORD STRAIGHT

1. "The Biggest Misconceptions People Have About Renewable Energy," *Wall Street Journal*, September 24, 2013, https://www.wsj.com/articles /the-biggest-misconceptions-people-have-about-renewable -energy-1380066859.

1. RENEWABLE ENERGY IN THE TWENTY-FIRST CENTURY

1. Sonia Smith, "Wind Power Capacity Has Surpassed Coal in Texas," *Texas Monthly*, December 2, 2017, https://www.texasmonthly.com/energy /wind-power-capacity-surpassed-coal-texas/.
2. Katrin Bennhold, "For First Time Since 1800s, Britain Goes a Day Without Burning Coal for Electricity," *New York Times*, April 21, 2017, https://www.nytimes.com/2017/04/21/world/europe/britain-burning -coal-electricity.html.
3. Anthony Dipaola, "Saudi Arabia Gets Cheapest Bids for Solar Power in Auction," *Bloomberg*, October 3, 2017, https://www.bloomberg.com /news/articles/2017-10-03/saudi-arabia-gets-cheapest-ever-bids-for -solar-power-in-auction.

2. ENERGY TRANSITIONS: FIRE TO ELECTRICITY

1. John Aberth, *An Environmental History of the Middle Ages: The Crucible of Nature* (London: Routledge, 2013).

2. The staff of Environmental Decision Making, Science, and Technology, "History of the Energy System," Carnegie Mellon University, accessed November 1, 2017, http://environ.andrew.cmu.edu/m3/s3/01history.shtml.

3. John U. Nef, "An Early Energy Crisis and Its Consequences," *Scientific American*, November 1, 1977, https://nature.berkeley.edu/er100/readings /Nef_1977.pdf, 140.

4. Nef, "Early Energy Crisis," 142.

5. Peter Brimblecombe, *The Big Smoke: A History of Air Pollution in London Since Medieval Times* (London: Methuen, 1987).

6. Ralph Waldo Emerson, *The Prose Works of Ralph Waldo Emerson, Volume 1* (Boston: James R. Osgood & Company, 1875), 362.

7. Peter A. O'Connor and Cutler J. Cleveland, "U.S. Energy Transitions, 1780–2010," *Energies* 7, no. 12 (November 2014): 7955–7993, http://www .mdpi.com/1996-1073/7/12/7955/htm.

8. Richard Rhodes, "Energy Transitions: A Curious History" (speech, Stanford University, September 19, 2007), Center for International Security and Cooperation, https://cisac.fsi.stanford.edu/sites/default /files/Rhodes-Energy_Transitions.pdf.

9. Brimblecombe, *Big Smoke*, 30.

10. O'Connor and Cleveland, *Energies* 7 (2014), "U.S. Energy Transitions 1780–2010," 7968.

11. O'Connor and Cleveland, "U.S. Energy Transitions," 7968.

12. O'Connor and Cleveland.

13. U.S. Energy Information Administration, "Energy Sources Have Changed Throughout the History of the United States," July 3, 2013, https://www.eia.gov/todayinenergy/detail.php?id=11951.

14. Martin A. Uman, "Why Did Benjamin Franklin Fly the Kite?," in *All About Lightning* (New York: Dover, 1986), http://ira.usf.edu/CAM /exhibitions/1998_12_McCollum/supplemental_didactics/23.Uman1 .pdf.

15. "Pearl Street Station," Engineering and Technology History Wiki, last modified November 23, 2017, http://ethw.org/Pearl_Street_Station.

16. Reuben Hull, "Electric Light by Water," *ASCE News*, October 6, 2015, http://news.asce.org/electric-light-by-water/.

17. *Power* is measured in watts, expressed in kilowatts (kW), megawatts (MW), and gigawatts (GW). 1 GW = 1,000 MW = 1,000,000 KW. For example, a home with ten lamps, each with a 100-watt lightbulb,

requires 1 kW of power for all the lights to be on at the same time. The average American home requires about 5 kW of power to run all appliances, lights, air conditioning, etc.

Electricity is measured in watts used per hour, expressed in kilowatt hours (kWh), megawatt hours (MWh), and gigawatt hours (GWh). 1 GWh = 1,000 MWh = 1,000,000 KWh. For example, a home that uses ten lamps, each with a 100-watt lightbulb, for four hours every evening will use 4 kWh of electricity every day. The average American home uses 10,766 kWh of electricity per year to run all appliances, lights, air conditioning, etc.

18. U.S. Energy Information Administration, "Competition Among Fuels for Power Generation Driven by Changes in Fuel Prices," July 13, 2012, https://www.eia.gov/todayinenergy/detail.php?id=7090.

19. Michelle L. Bell, Devra L. Davis, and Tony Fletcher, "A Retrospective Assessment of Mortality from the London Smog Episode of 1952: The Role of Influenza and Pollution," *Environmental Health Perspectives* 112, no. 1 (January 2004): 6–8, https://www.ncbi.nlm.nih.gov/pmc/articles/PMC1241789/.

20. Discounting is a process used to calculate the present value of a stream of future cash flows by accounting for the time value of money. The cost of capital is the cost of financing projects using both debt and equity capital, calculated as the Weighted Average Cost of Capital (WACC). Refer to appendix A for the calculation of WACC.

21. Dwight D. Eisenhower, "Atoms for Peace" (speech, 470th Plenary Meeting of the United Nations General Assembly, December 8, 1953), available at https://www.iaea.org/about/history/atoms-for-peace-speech.

22. U.S. Department of Energy, *The History of Nuclear Energy*, DOE/NE-0088, https://www.energy.gov/sites/prod/files/The%20History%20of%20Nuclear%20Energy_0.pdf.

23. IEA Statistics, "Electricity Production from Nuclear Sources (% of Total)," World Bank, 2014, http://data.worldbank.org/indicator/EG.ELC.NUCL.ZS?locations=US.

24. U.S. Department of Energy, *History of Nuclear Energy*.

25. Associated Press, "14-Year Cleanup at Three Mile Island Concludes," *New York Times*, August 15, 1993, http://www.nytimes.com/1993/08/15/us/14-year-cleanup-at-three-mile-island-concludes.html.

26. IEA Statistics, "Electricity Production from Nuclear Sources."

27. Peter Fairley, "Why Don't We Have More Nuclear Power?," *MIT Technology Review*, May 28, 2015, https://www.technologyreview.com /s/537816/why-dont-we-have-more-nuclear-power/.

28. Jim Polson, "More Than Half of America's Nuclear Reactors Are Losing Money," *Bloomberg*, June 14, 2017, https://www.bloomberg.com/news /articles/2017-06-14/half-of-america-s-nuclear-power-plants-seen-as -money-losers.

29. Robert C. Armstrong et al., "The Frontiers of Energy," *Nature Energy* 1, no. 11 (January 2016).

30. IEA Statistics, "Electricity Production from Nuclear Sources."

31. Oliver Kuhn, "Ancient Chinese Drilling," *Recorder* 29, no. 6 (June 2004), http://csegrecorder.com/articles/view/ancient-chinese-drilling.

32. U.S. Department of Energy, "Fossil Energy Study Guide: Natural Gas," https://www.energy.gov/sites/prod/files/2017/05/f34/MS_NatGas _Studyguide.pdf.

33. Mark Axford, "Gas Turbine Orders: 1998 Boom, 2001 Bust, 2004 Rebound?," Energy Central, January 22, 2004, http://www.energycentral .com/c/gn/gas-turbine-orders-1998-boom-2001-bust-2004-rebound.

34. Bruce A. Wells and Kris Wells, "Shooters—A 'Fracking' History," American Oil and Gas Historical Society, May 7, 2017, https://aoghs .org/technology/hydraulic-fracturing/.

35. "The Father of Fracking," *Schumpeter* (blog), *Economist*, August 3, 2013, https://www.economist.com/news/business/21582482-few-business people-have-done-much-change-world-george-mitchell-father.

36. U.S. Energy Information Administration, "Competition Among Fuels."

37. U.S. Energy Information Administration, "Competition Among Fuels."

38. U.S. Energy Information Administration, "Competition Among Fuels."

3. THE RISE OF RENEWABLES

1. "2015 Renewable Energy Investments Were Double Fossil Fuel Power Plant Investments," CleanTechnica, March 26, 2016, https://clean technica.com/2016/03/26/2015-renewable-energy-investments-were -double-fossil-fuel-power-plant-investments/.

2. Emily Gosden, "Global Renewable Power Capacity Overtakes Coal as 500,000 Solar Panels Installed Every Day," *Telegraph*, October 26, 2016,

http://www.telegraph.co.uk/business/2016/10/25/global-renewable
-power-capacity-overtakes-coal-as-500000-solar-p/.

3. Bloomberg New Energy Finance, *New Energy Outlook 2017*, last
modified June 27, 2017, https://about.bnef.com/new-energy-outlook/.

4. "Some Facts About the Three Gorges Project," Embassy of the People's
Republic of China in the United States of America, accessed August 12,
2017, http://www.china-embassy.org/eng/zt/sxgc/t36512.htm.

5. "Hydropower," International Energy Agency, accessed June 17, 2017,
https://www.iea.org/topics/renewables/hydropower/.

6. IEA Statistics, "Electricity Production from Hydroelectric Sources
(% of Total)," World Bank, 2014, http://data.worldbank.org/indicator
/EG.ELC.HYRO.ZS.

7. IEA Statistics, "Electricity Production from Hydroelectric Sources."

8. Dan Drollette, "Energy from the Motion of the Ocean: A Former
Surfer Designs a Buoy that Can Convert Wave Motion into
Electricity," *Fortune Small Business*, December 15, 2006, http://money
.cnn.com/2006/12/14/magazines/fsb/nextlittlething_wave_power.fsb
/index.htm.

9. "Energy Use in Sweden," sweden.se, last modified January 12, 2018,
https://sweden.se/society/energy-use-in-sweden/.

10. "Bioenergy (Biofuels and Biomass)," Environmental and Energy
Study Institute, last modified October 23, 2008, http://www.eesi.org
/topics/bioenergy-biofuels-biomass/description.

11. Melissa C. Lott, "The U.S. Now Uses More Corn for Fuel Than for Feed,"
Scientific American, October 7, 2011, https://blogs.scientificamerican.com
/plugged-in/the-u-s-now-uses-more-corn-for-fuel-than-for-feed/.

12. Tim Searchinger and Ralph Heimlich, "Avoiding Bioenergy Competition
for Food Crops and Land: Creating a Sustainable Food Future, Installment
Nine," World Resources Institute, January 29, 2015, http://www.wri.org
/publication/avoiding-bioenergy-competition-food-crops-and-land.

13. "Geothermal Power," BP, last modified June 13, 2017, http://www
.bp.com/en/global/corporate/energy-economics/statistical-review-of
-world-energy/renewable-energy/geothermal-power.html.

14. IRENA "Renewable Power Generation Costs in 2017," http://www
.irena.org/-/media/Files/IRENA/Agency/Publication/2018/Jan
/IRENA_2017_Power_Costs_2018.pdf.

4. RENEWABLE WIND ENERGY

1. Jan Hylleberg et al., "Denmark—Wind Power Hub: Profile of the Danish Wind Industry," Danish Wind Industry Association, 2008, http://www.windpower.org/download/378/profilbrochure_2008pdf.

2. Lester R. Brown, *The Great Transition: Shifting from Fossil Fuels to Solar and Wind Energy* (New York: Norton, 2015), 86.

3. Magdi Ragheb, "Historical Wind Generators Machines," mragheb .com, February 6, 2013, http://mragheb.com/NPRE%20475%20Wind %20Power%20Systems/Historical%20Wind%20Generators%20 Machines.pdf.

4. Dan Ancona and Jim McVeigh, "Wind Turbine—Materials and Manufacturing Fact Sheet," CiteSeerX, August 29, 2001, http://citeseerx .ist.psu.edu/viewdoc/download?doi=10.1.1.464.5842&rep=rep1&type=pdf.

5. Doubling wind speed → 2^3 → 8× energy output. For example, a wind turbine sited in a location with average wind speeds of 20 mph will generate 800 percent as much electricity as a wind turbine sited with average wind speeds of 10 mph. Conversely, a reduction in wind speed will dramatically reduce energy output. For example, a wind turbine that is poorly sited and experiences 80 percent of the wind that it was designed for will generate only 51 percent of expected electricity.

6. Zachary Shahan, "Wind Power Awesomeness," CleanTechnica, October 11, 2013, https://cleantechnica.com/2013/10/11/wind-power -awesomeness/.

7. "New York Wind Energy Guide for Local Decision Makers," New York State Energy Research and Development Authority, accessed June 5, 2017, https://www.nyserda.ny.gov/Researchers-and-Policymakers/Power -Generation/Wind/Large-Wind/New-York-Wind-Energy-Guide -Local-Decision-Makers.

8. U.S. Energy Information Administration, "Table 6.7.B. Capacity Factors for Utility Scale Generators Not Primarily Using Fossil Fuels, January 2013–January 2018," *Electric Power Monthly*, last modified March 23, 2018, https://www.eia.gov/electricity/monthly/epm_table _grapher.php?t=epmt_6_07_b.

9. Electricity generated = power rating × capacity factor × hours operated. In this example, electricity generated = 5 MW × 35 percent × 24 hours/ day × 365 days/year = 15,330 MWh/year.

10. "Report on Wind Turbine Gearbox and Direct-Drive Systems Out Now," *Offshore Wind*, September 19, 2014, http://www.offshorewind .biz/2014/09/19/report-on-wind-turbine-gearbox-and-direct-drive -systems-out-now/.

11. Lazard, "Levelized Cost of Energy 2017," November 2, 2017, https:// www.lazard.com/perspective/levelized-cost-of-energy-2017/.

12. Annie Sneed, "Moore's Law Keeps Going, Defying Expectations," *Scientific American*, May 19, 2015, https://www.scientificamerican.com /article/moore-s-law-keeps-going-defying-expectations/.

13. 1 GW = 1,000 MW.

14. "Renewables 2017 Global Status Report," REN21, March 13, 2017, http://www.ren21.net/wp-content/uploads/2017/06/17-8399 _GSR_2017_Full_Report_0621_Opt.pdf.

15. *Lazard's Levelized Cost of Energy Analysis—Version 10.0*, December 2016, https://www.lazard.com/media/438038/levelized-cost-of-energy-v100 .pdf.

16. Ryan Wiser et al., "Expert Elicitation Survey on Future Wind Energy Costs," *Nature Energy* 1, no. 10 (September 12, 2016), https://www .nature.com/articles/nenergy2016135.epdf?author_access_token =xOjt15xAsgbwf-DTbC9umtRgNojAjWel9jnR3ZoTvoPmotc EncNIRUyqt3vi2Zdm55gFQx3FMImKGoGh8VsPowqN8Ae ZekJAOtf6AfxskkGU8raC7OZ5Y_20S7qTMDRvAjSHfuoi9oAte8h 3yQ3nDw%3D%3D.

17. Eric Rosenbloom, "Areas of Industrial Wind Facilities," AWEO.org, last modified February 17, 2017, http://www.aweo.org/windarea.html.

18. Jennifer Oldham, "Wind Is the New Corn for Struggling Farmers," *Bloomberg*, October 6, 2016, https://www.bloomberg.com/news/articles /2016-10-06/wind-is-the-new-corn-for-struggling-farmers.

19. Gary Schnitkey and the Department of Agricultural and Consumer Economics, University of Illinois, *Revenue and Costs for Corn, Soybeans, Wheat, and Double-Crop Soybeans, Actual for 2011 Through 2016, Projected 2017 and 2018*, February 2018, http://www.farmdoc.illinois.edu/manage /actual_projected_costs.pdf.

20. Oldham, "Wind Is the New Corn."

21. Oldham, "Wind Is the New Corn."

22. Hill Country Wind Power, "Wind Basics," last modified August 30, 2010, http://www.hillcountrywindpower.com/wind-basics.php.

23. Jan Dell and Matthew Klippenstein, "Wind Power Could Blow Past Hydro's Capacity Factor by 2020," *Greentech Media*, February 8, 2017, https://www.greentechmedia.com/articles/read/wind-power-could -blow-past-hydros-capacity-factor-by-2020.

24. John Alessi, "The Battle For Cape Wind: An Analysis of Massachusetts Newspapers and Their Framing of Offshore Wind Energy," p11.https://www.capewind.org/article/2002/10/03/1003-new-poll-finds -strong-public-support-cape-wind-project.

25. John Leaning, "Cronkite Spins Ad for Foes of Wind Farm," *Cape Cod Times*, January 30, 2003, last modified January 5, 2011, http://www .capecodtimes.com/article/20030130/NEWS01/301309987.

26. Jason Samenow, "Blowing Hard: The Windiest Time of Year and Other Fun Facts on Wind," *Washington Post*, updated March 31, 2016, https://www.washingtonpost.com/news/capital-weather-gang/wp /2014/03/26/what-are-the-windiest-states-and-cities-what-is-d-c-s -windiest-month/?utm_term=.4ac52623e129.

27. "Wind in Numbers," Global Wind Energy Council, last modified May 5, 2017, http://www.gwec.net/global-figures/wind-in-numbers/.

5. RENEWABLE SOLAR ENERGY

1. Jeff Tsao, Nate Lewis, and George Crabtree, "Solar FAQs," Sandia National Laboratories, April 20, 2006, http://www.sandia.gov/~jytsao /Solar%20FAQs.pdf.

2. LaMar Alexander, *Off the Grid: Simple Solar Homesteading*, (Lulu, 2011), 112.

3. Jeremy Hsu, "Vanguard 1, First Solar-Powered Satellite, Still Flying at 50," Space.com, March 18, 2008, https://www.space.com/5137-solar -powered-satellite-flying-50.html.

4. Megan Geuss, "Japanese Company Develops a Solar Cell with Record-Breaking 26%+ Efficiency," Ars Technica, March 22, 2017, https:// arstechnica.com/science/2017/03/japanese-company-develops-a-solar -cell-with-record-breaking-26-efficiency/.

5. Cost per watt = manufacturing cost / rated output. $600 / 200 watts = $3 per watt.

6. Ran Fu et al., *U.S. Solar Photovoltaic System Cost Benchmark: Q1, 2016*, National Renewable Energy Laboratory, September 28, 2016, https:// www.nrel.gov/docs/fy16osti/66532.pdf.

7. Ramez Naam, "Smaller, Cheaper, Faster: Does Moore's Law Apply to Solar Cells?," *Scientific American*, March 16, 2011, https://blogs.scientific american.com/guest-blog/smaller-cheaper-faster-does-moores-law -apply-to-solar-cells/.

8. Chris Martin, "Solar Panels Now So Cheap Manufacturers Probably Selling at Loss," *Bloomberg*, December 30, 2016, https://www.bloomberg .com/news/articles/2016-12-30/solar-panels-now-so-cheap-manufacturers -probably-selling-at-loss.

9. Geoffrey Carr, "Sunny Uplands," *Economist*, November 21, 2012, https://www.economist.com/news/21566414-alternative-energy-will -no-longer-be-alternative-sunny-uplands.

10. Edward S. Rubin, Inês M. L. Azevedo, Paulina Jaramillo, and Sonia Yeh, "A Review of Learning Rates for Electricity Supply Technologies," *Energy Policy* 86 (2015), 198.

11. Solar Energy Industries Association, "Solar Photovoltaic Technology— Presentation Transcript," March 15, 2009, http://www.seia.org/research -resources/solar-photovoltaic-technology.

12. U.S. Energy Information Administration, "More Than Half of Small-Scale Photovoltaic Generation Comes from Residential Rooftops," June 1, 2017, https://www.eia.gov/todayinenergy/detail.php?id=31452.

13. Solar Energy Industries Association, *Solar Means Business 2016 Report*, October 19, 2016, http://www2.seia.org/l/139231/2016-10-18/s9lzt.

14. U.S. Energy Information Administration, "Small-Scale Photovoltaic Generation."

15. Renewable Energy Policy Network for the 21st Century, *Renewables 2017 Global Status Report*, March 13, 2017, http://www.ren21.net/wp -content/uploads/2017/06/17-8399_GSR_2017_Full_Report_0621 _Opt.pdf.

16. Simon Stevens and Kevin Smith, "Is CSP an Expensive or a Viable Investment?," New Energy Update, April 17, 2015, http://analysis.newenergy update.com/csp-today/markets/csp-expensive-or-viable-investment.

17. Lazard, "Levelized Cost of Energy Analysis 10.0," December 15, 2016, https://www.lazard.com/perspective/levelized-cost-of-energy-analysis -100/.

18. Solar Energy Industries Association, "Solar Industry Research Data," last modified March 16, 2018, http://www.seia.org/research-resources /solar-industry-data.

6. FINANCING RENEWABLE ENERGY

1. Meredith Fowlie, "The Renewable Energy Auction Revolution," *Energy Institute at Haas* (blog), August 7, 2017, https://energyathaas. wordpress.com/2017/08/07/the-renewable-energy-auction-revolution/.

2. Galen Barbose, *U.S. Renewables Portfolio Standards: 2017 Annual Status Report*, Lawrence Berkeley National Laboratory, July 21, 2017, https:// emp.lbl.gov/sites/default/files/2017-annual-rps-summary-report.pdf.

3. United Nations, *United Nations Framework Convention on Climate Change*, May 9, 1992, available at http://unfccc.int/files/essential _background/background_publications_htmlpdf/application/pdf /conveng.pdf.

4. Raphael Calel, "Climate Change and Carbon Markets: A Panoramic History," Centre for Climate Change Economics and Policy Working Paper No. 62 and Grantham Research Institute on Climate Change and the Environment Working Paper No. 52, July 12, 2011, http://eprints.lse .ac.uk/37397/1/Climate_change_and_carbon_markets_a_panoramic _history(author).pdf.

5. Lauraine G. Chestnut and David M. Mills, "A Fresh Look at the Benefits and Costs of the US Acid Rain Program," *Journal of Environmental Management* 77, no. 3 (December 2005): 252–266, https://cfpub .epa.gov/si/si_public_record_report.cfm?dirEntryID=139587.

6. Institute for Energy Research, "China's Renewable Industry Still Getting CDM-Funded Projects," July 23, 2012, https://instituteforenergyresearch .org/analysis/chinas-renewable-industry-still-getting-cdm-funded -projects/.

7. California Air Resources Board, "Cap-and-Trade Program," CA.gov, last modified March 30, 2018, https://www.arb.ca.gov/cc/capandtrade /capandtrade.htm.

8. Debra Kahn, "China Is Preparing to Launch the World's Biggest Carbon Market," *Scientific American*, August 14, 2017, https://www .scientificamerican.com/article/china-is-preparing-to-launch-the -world-rsquo-s-biggest-carbon-market/.

9. Benjamin Esty, Suzie Harris, and Kathy Krueger, "An Overview of the Project Finance Market," Harvard Business School, December 13, 1999, http://www.austraclear.net/wp-content/uploads/2016/10/Harvard-Business -School-1999-An-Overview-of-the-Project-Finance-Model.pdf.

10. *Lazard's Levelized Cost of Energy Analysis—Version 10.0*, December 2016, https://www.lazard.com/media/438038/levelized-cost-of-energy-v100 .pdf.

11. FS-UNEP Collaborating Centre for Climate and Sustainable Energy Finance, "Global Trends in Renewable Energy Investment 2017," January 10, 2017, http://fs-unep-centre.org/sites/default/files/publications /globaltrendsinrenewableenergyinvestment2017.pdf.

12. FS-UNEP Collaborating Centre, "Global Trends"; "Masdar Achieves Financial Closure of London Array Project," *Gulf News*, October 10, 2013, http://gulfnews.com/business/sectors/investment/masdar-achieves -financial-closure-of-london-array-project-1.1241833.

13. John Pavlus et al., "World Changing Ideas 2010," *Scientific American*, December 15, 2010, https://www.scientificamerican.com/article/world -changing-ideas-dec10/.

14. Herman K. Trabish, "Why Solar Financing Is Moving from Leases to Loans," Utility Dive, August 17, 2015, http://www.utilitydive.com/news /why-solar-financing-is-moving-from-leases-to-loans/403678/.

15. Trabish, "Why Solar Financing Is Moving."

16. Katherine Tweed, "Pay-As-You-Go Transactions in Off-Grid Solar Top $41M in Late 2016," Greentech Media, May 19, 2017, https://www .greentechmedia.com/articles/read/off-grid-solar-pay-as-you-go -transactions-top-41m-in-late-2016.

17. Abraham Louw, *Clean Energy Investment Trends, 3Q 2017*, Bloomberg New Energy Finance, October 5, 2017, https://data.bloomberglp.com /bnef/sites/14/2017/10/BNEF-Clean-Energy-Investment-Trends-3Q -2017.pdf.

18. Stefan Nicola and Marc Roca, "Solar Returns Declining as Investor Interest Seen Rising," *Bloomberg*, June 12, 2014, https://www.bloomberg .com/news/articles/2014-06-12/solar-returns-declining-as-investor -interest-seen-rising.

7. ENERGY TRANSITIONS: OATS TO OIL

1. Ben Johnson, "The Great Horse Manure Crisis of 1894," *Historic UK History Magazine*, December 2, 2013, http://www.historic-uk.com /HistoryUK/HistoryofBritain/Great-Horse-Manure-Crisis-of -1894/.

2. Dayville Hay & Grain Inc., "Calorie Requirements for Horses," accessed July 2, 2017, http://www.dayvillesupply.com/hay-and-horse -feed/calorie-needs.html.

3. Brian Groom, "The Wisdom of Horse Manure," *Financial Times*, September 2, 2013, https://www.ft.com/content/238b1038-13bb-11e3-9289 -00144feabdco.

4. "Internal Combustion Engine," New World Encyclopedia, last modified March 4, 2018, http://www.newworldencyclopedia.org/entry/Internal _combustion_engine.

5. *Encyclopaedia Britannica*, s.v. "Karl Benz," last modified March 29, 2018, https://www.britannica.com/biography/Karl-Benz.

6. Martin V. Melosi, "The Automobile and the Environment in American History," Automobile in American Life and Society, University of Michigan-Dearborn, September 9, 2005, http://www.autolife.umd .umich.edu/Environment/E_Overview/E_Overview3.htm.

7. Eric J. Dahl, "Naval Innovation: From Coal to Oil," National Defense University, 2001, http://www.dtic.mil/dtic/tr/fulltext/u2/a524799.pdf.

8. Dahl, "Naval Innovation."

9. Bruce A. Wells, "Petroleum and Sea Power," American Oil and Gas Historical Society, June 9, 2014, http://aoghs.org/petroleum-in-war /petroleum-and-sea-power/.

10. Stacy C. Davis, Susan E. Williams, and Robert G. Boundy, "Energy," in *Transportation Energy Data Book*, 35th ed. (Oak Ridge, Tenn.: Oak Ridge National Laboratory, 2016), https://info.ornl.gov/sites/publications/Files /Pub69643.pdf.

11. Peter A. O'Connor and Cutler J. Cleveland, "U.S. Energy Transitions, 1780–2010," *Energies* 7, no. 12 (November 2014): 7955–7993, http://www .mdpi.com/1996-1073/7/12/7955/htm.

12. Jonathan M. Harris and Brian Roach, "Energy: The Great Transition," in *Environmental and Natural Resource Economics: A Contemporary Approach*, 4th ed. (Medford, Mass.: Tufts University, 2016), http://www .ase.tufts.edu/gdae/Pubs/te/ENRE/4/Ch11_Energy_4E.pdf.

8. THE RISE OF ELECTRIC VEHICLES

1. Robert L. Bradley Jr., "Electric Vehicles: As in 1896, the Wrong Way to Go," Institute for Energy Research, October 19, 2010,

http://instituteforenergyresearch.org/analysis/electric-vehicles-as
-in-1896-the-wrong-way-to-go/.

2. Rosenblum Law Firm, "Who Got America's First Speeding Ticket?,"
   New York Speeding Ticket Fines, June 20, 2016, http://newyorkspeeding
   fines.com/americas-speeding-ticket/.

3. Jim Motavalli, "Porsche's Long-Buried First Vehicle Was an Electric
   Car, and It Was Built Back in 1898," Mother Nature Network, January 8,
   2014, https://www.mnn.com/green-tech/transportation/blogs/porsches
   -long-buried-first-vehicle-was-an-electric-car-and-it-was.

4. U.S. Department of Energy, www.fueleconomy.gov, "All-Electric
   Vehicles," https://www.fueleconomy.gov/feg/evtech.shtml.

5. Dan Strohl, "Ford, Edison and the Cheap EV that Almost Was,"
   Wired, June 18, 2010, https://www.wired.com/2010/06/henry-ford-thomas
   -edison-ev/.

6. Martin V. Melosi, "The Automobile and the Environment in American
   History," Automobile and the Environment in American History, University of Michigan-Dearborn, September 9, 2005, http://www.autolife
   .umd.umich.edu/Environment/E_Overview/E_Overview3.htm.

7. Bob Casey, "General Motors' EV1," The Henry Ford, June 22, 2015,
   https://www.thehenryford.org/explore/blog/general-motors-ev1/.

8. Elon Musk, "The Secret Tesla Motors Master Plan (Just Between
   You and Me)," Tesla, August 2, 2006, https://www.tesla.com/blog/secret
   -tesla-motors-master-plan-just-between-you-and-me.

9. George E. Blomgren, "The Development and Future of Lithium Ion
   Batteries," *Journal of the Electrochemical Society* 164, no. 1 (December 2016),
   http://jes.ecsdl.org/content/164/1/A5019.full.

10. Electric vehicles receive a miles-per-gallon (mpg) equivalent rating
    with which consumers can compare fuel efficiency against gasoline
    -powered vehicles.

11. Kim Reynolds, "2008 Tesla Roadster First Drive," *Motor Trend*,
    January 22, 2008, http://www.motortrend.com/cars/tesla/roadster/2008
    /2008-tesla-roadster/.

12. "Tesla Roadster Sport vs Model S," TwinRev, accessed July 31, 2017,
    http://twinrev.com/cars/Tesla-Roadster-Sport-vs-Tesla-Model-S.

13. Christopher DeMorro, "Cost of the Tesla Model E Exaggerated in
    Flawed Study," CleanTechnica, March 4, 2014, https://cleantechnica
    .com/2014/03/04/flawed-study-exaggerates-cost-tesla-model-e/.

14. "Federal Tax Credits for All-Electric and Plug-In Hybrid Vehicles," www.fueleconomy.gov, December 4, 2009, https://www.fueleconomy .gov/feg/taxevb.shtml.

15. Office of Energy Efficiency and Renewable Energy, "Electric Vehicles: Tax Credits and Other Incentives," September 16, 2015, https://energy.gov /eere/electricvehicles/electric-vehicles-tax-credits-and-other-incentives.

16. International Energy Agency, "Global EV Outlook 2017: Two Million and Counting," June 6, 2017, https://www.iea.org/publications/free publications/publication/GlobalEVOutlook2017.pdf.

17. Sarah J. Gerssen-Gondelach and André P. C. Faaij, "Performance of Batteries for Electric Vehicles on Short and Longer Term," *Journal of Power Sources* 212 (August 2012): 111–129.

18. Christophe Pillot, "Battery Market Development for Consumer Electronics, Automotive, and Industrial: Materials Requirements and Trends," Avicenne Energy (presentation, Qinghai EV Rally 2015, Xining, China, June 15–18, 2015), http://www.avem.fr/docs/pdf/Avicenne DiapoXining.pdf.

19. Leslie Shaffer, "JPMorgan Thinks the Electric Vehicle Revolution Will Create a Lot of Losers," *CNBC*, August 22, 2017, https://www.cnbc .com/2017/08/22/jpmorgan-thinks-the-electric-vehicle-revolution -will-create-a-lot-of-losers.html.

20. Steven Szakaly and Patrick Manzi, *NADA Data, 2015: Annual Financial Profile of America's Franchised New-Car Dealerships*, December 6, 2016, https://www.nada.org/WorkArea/DownloadAsset.aspx?id =21474839497.

21. Rob Wile, "Credit Suisse Gives Point-by-Point Breakdown Why Tesla Is Better than Your Regular Car," *Business Insider*, August 14, 2014, http://www.businessinsider.com/credit-suisse-on-tesla-2014-8.

22. Tibor Blomhäll, "Test Drive of a Petrol Car," Tesla Club Sweden, April 22, 2015, http://teslaclubsweden.se/test-drive-of-a-petrol-car/.

23. Thomas Fisher, "Will Tesla Alone Double Global Demand for Its Battery Cells?," Green Car Reports, September 3, 2013, http://www.greencarre ports.com/news/1086674_will-tesla-alone-double-global-demand -for-its-battery-cells/page-2.

24. "Battery Cell Production Begins at the Gigafactory," Tesla, January 4, 2017, https://www.tesla.com/blog/battery-cell-production-begins-gigafactory.

25. Fred Lambert "Tesla Is Now Claiming 35% Battery Cost Reduction at Gigafactory 1," Electrek, February 18, 2017, https://electrek.co/2017/02/18/tesla-battery-cost-gigafactory-model-3/.

26. Elon Musk, "The Future We're Building—and Boring" (TED Talk, TED2017, Vancouver, BC, April 24–28, 2017), https://www.ted.com/talks/elon_musk_the_future_we_re_building_and_boring/transcript.

27. U.S. Department of Energy, "Maps and Data," Alternative Fuels Data Center, last updated March 2018, https://www.afdc.energy.gov/data/categories/vehicles.

28. U.S. Energy Information Administration, "Table 5.6.A. Average Price of Electricity to Ultimate Customers by End-Use Sector," *Electric Power Monthly*, March 23, 2018, https://www.eia.gov/electricity/monthly/epm_table_grapher.php?t=epmt_5_6_a.

29. Idaho National Laboratory, "Comparing Energy Costs per Mile for Electric and Gasoline-Fueled Vehicles," June 25, 2011, https://avt.inl.gov/sites/default/files/pdf/fsev/costs.pdf.

30. "Monthly Plug-In Sales Scorecard," InsideEVs, accessed July 10, 2017, http://insideevs.com/monthly-plug-in-sales-scorecard/.

31. Nikki Gordon-Bloomfield, "95 Percent of All Trips Could Be Made in Electric Cars, Says Study," Green Car Reports, January 13, 2012, http://www.greencarreports.com/news/1071688_95-of-all-trips-could-be-made-in-electric-cars-says-study.

32. Lauren Tyler, "Report: Public Charging Remains No. 1 Concern for EV Drivers," *NGT News—Next-Gen Transportation*, July 10, 2017, https://ngtnews.com/report-ev-drivers-still-concerned-public-charging-availability.

33. "Gas Station Industry Statistics," StatisticsBrain, September 3, 2016, https://www.statisticbrain.com/gas-station-statistics/.

34. "Alternative Fueling Station Locator," Alternative Fuels Data Center, accessed July 11, 2017, https://www.afdc.energy.gov/locator/stations/results?fuel=ELEC.

35. Patrick Sisson, "Ford Announces Details of $4.5 Billion Investment in Electric Vehicles," Curbed, January 3, 2017, https://www.curbed.com/2017/1/3/14153954/ford-electric-vehicle-flat-rock-autonomous-car.

9. PARITY

1. Kathy Finn, "Solar Leasing Widens the Appeal of Sun Power," *New Orleans Advocate*, September 23, 2014, http://www.theadvocate. com/new_orleans/news/business/article_3c5b46f0-fe07-5eb7-afd7 -84ce3ef12d9f.html.

2. Meghan French Dunbar, "SolarCity Is Transforming the Renewable Energy Industry One Rooftop at a Time," *Conscious Company Media*, January 5, 2016, https://consciouscompanymedia.com/the-new -economy/solarcity-is-transforming-the-renewable-energy-industry -one-rooftop-at-a-time/.

3. Elizabeth Bast et al., *Empty Promises: G20 Subsidies to Oil, Gas and Coal Production*, Oil Change International, February 11, 2015, http:// priceofoil.org/content/uploads/2015/11/Empty-promises_main-report .2015.pdf.

4. Bast et al., *Empty Promises*.

5. "TrueCapture," NEXTracker, July 6, 2017, https://www.nextracker.com /product-services/truecapture/.

6. Scott Moskowitz, "The Global PV Tracker Landscape 2016: Prices, Forecasts, Market Shares and Vendor Profiles," GTM Research, accessed July 15, 2017, https://www.greentechmedia.com/research/report /the-global-pv-tracker-landscape-2016.

7. Office of Energy Efficiency and Renewable Energy (EERE), "Next-Generation Wind Technology," February 8, 2018, https://energy.gov /eere/next-generation-wind-technology.

8. EERE, "Next-Generation Wind Technology."

9. Katherine Tweed, "Survey: 76 Percent of Consumers Don't Trust Their Utility," Greentech Media, July 8, 2013, https://www.greentechmedia .com/articles/read/consumer-trust-in-utilities-continues-to-nosedive.

10. Bryan Bollinger and Kenneth Gillingham, "Peer Effects in the Diffusion of Solar Photovoltaic Panels," *Marketing Science* 31, no. 6 (September 2012): 900–912.

11. Sharon O'Malley, "Here Comes the Sun: More Builders Offer Solar Arrays as Option on Homes," Construction Dive, April 27, 2015, http:// www.constructiondive.com/news/here-comes-the-sun-more-builders -offer-solar-arrays-as-option-on-homes/391356/.

12. Sophie Vorrath, "One-Quarter of Australian Homes Now Have Solar," *pv magazine*, July 6, 2017, https://www.pv-magazine.com/2017/07/06/one-quarter-of-australian-homes-now-have-solar/.

13. Daniel Silkstone, "The Suburbs Where Renewables Rule (Is Yours on the List?)," ARENA Wire, September 1, 2017, https://arena.gov.au/blog/climatecouncil/.

14. Solar Energy Industries Association, *Solar Market Insight Report, 2016 Q3*, September 12, 2016, https://www.seia.org/research-resources/solar-market-insight-report-2016-q3.

15. Chris Baraniuk, "Why Apple and Google Are Moving into Solar Energy," *BBC Future*, October 14, 2016, http://www.bbc.com/future/story/20161013-why-apple-and-google-are-going-solar.

16. Doug McMillon, "Walmart Offers New Vision for the Company's Role in Society," Walmart, November 4, 2016, http://news.walmart.com/2016/11/04/walmart-offers-new-vision-for-the-companys-role-in-society.

17. Lester R. Brown, *The Great Transition: Shifting from Fossil Fuels to Solar and Wind Energy* (New York: Norton, 2015), 145.

18. Chris Martin, "U.S. Solar Surged 95 Percent to Become Largest Source of New Energy," *Bloomberg*, February 15, 2017, https://www.bloomberg.com/news/articles/2017-02-15/u-s-solar-surged-95-to-become-largest-source-of-new-energy.

19. U.S. Energy Information Administration, "Renewable Generation Capacity Expected to Account for Most 2016 Capacity Additions," January 10, 2017, https://www.eia.gov/todayinenergy/detail.php?id=29492.

20. Michael Safi, "Indian Solar Power Prices Hit Record Low, Undercutting Fossil Fuels," *Guardian*, May 10, 2017, https://www.theguardian.com/environment/2017/may/10/indian-solar-power-prices-hit-record-low-undercutting-fossil-fuels.

21. Saurabh Mahapatra, "New Low Solar Price Record Set in Chile—2.91¢ Per kWh!," CleanTechnica, August 18, 2016, https://cleantechnica.com/2016/08/18/new-low-solar-price-record-set-chile-2-91%C2%A2-per-kwh/.

22. Saurabh Mahapatra, "JinkoSolar, Marubeni Score Lowest-Ever Solar PV Bid at 2.42¢/kWh in Abu Dhabi," CleanTechies, September 20, 2016,

http://cleantechies.com/2016/09/20/jinkosolar-marubeni-score-lowest
-ever-solar-pv-at-us%C2%A22-42kwh-in-abu-dhabi/.

23. Jason Deign, "India's Record-Low Wind and Solar Prices May Not Be Sustainable," Greentech Media, September 25, 2017, https://www.greentechmedia.com/articles/read/indias-renewable-energy-auctions-may-not-be-sustainable.

24. Office of Energy Efficiency and Renewable Energy, "Next-Generation Wind Technology," February 8, 2018, https://energy.gov/eere/next-generation-wind-technology.

25. Brown, *The Great Transition*, 94.

26. Berkshire Hathaway Energy, "Renewables," April 14, 2016, https://www.berkshirehathawayenergyco.com/environment/renewables.

27. Evelyn Cheng, "Warren Buffet Says He's Got a 'Big Appetite' for a Solar or Wind Project," *CNBC*, May 6, 2017, https://www.cnbc.com/2017/05/06/warren-buffett-says-hes-got-a-big-appetite-for-a-solar-or-wind-project.html.

28. Jess Shankleman, "BlackRock Busts $1 Billion Green Power Goal with Second Fund," *Bloomberg*, July 5, 2017, https://www.bloomberg.com/news/articles/2017-07-05/blackrock-busts-1-billion-green-power-goal-with-second-fund.

29. Bloomberg New Energy Finance, *Clean Energy Investment—3Q 2017 Trends*, October 5, 2017, https://about.bnef.com/blog/clean-energy-investment-3q-2017-trends/.

30. International Energy Agency, "Global Energy Investment Fell for a Second Year in 2016 as Oil and Gas Spending Continues to Drop," July 11, 2017, https://www.iea.org/newsroom/news/2017/july/global-energy-investment-fell-for-a-second-year-in-2016-as-oil-and-gas-spending-c.html.

31. Neil Winton, "Electric Car Price Parity Expected Next Year—Report," *Forbes*, May 22, 2017, https://www.forbes.com/sites/neilwinton/2017/05/22/electric-car-price-parity-expected-next-year-report/#712796a47922.

32. Jess Shankleman, "Pretty Soon Electric Cars Will Cost Less than Gasoline," *Bloomberg*, May 26, 2017, https://www.bloomberg.com/news/articles/2017-05-26/electric-cars-seen-cheaper-than-gasoline-models-within-a-decade; Fred Lambert, "Automakers Need to Brace for the

Impact of 1 Billion Electric Vehicles," Electrek, September 5, 2017, https://electrek.co/2017/09/05/automakers-1-billion-electric-vehicles/.

33. KBB.com editors, "Class of 2017: New Cars Ready to Roll," Kelley Blue Book, November 10, 2016, https://www.kbb.com/car-news/all-the -latest/class-of-2017-new-cars-ready-to-roll/2100000298/.

34. Robert Rapier, "U.S. Electric Vehicle Sales Soared in 2016," *Forbes*, February 5, 2017, https://www.forbes.com/sites/rrapier/2017/02/05/u-s -electric-vehicle-sales-soared-in-2016/#75c9ba0c217f.

35. Don Sherman, "How a Car Is Made: Every Step from Invention to Launch," *Car and Driver*, November 18, 2015, https://blog.caranddriver .com/how-a-car-is-made-every-step-from-invention-to-launch/.

36. "Volvo Cars to Go All Electric," Volvo Car Group Global Newsroom, July 5, 2017, https://www.media.volvocars.com/global/en-gb/media /pressreleases/210058/volvo-cars-to-go-all-electric.

37. Darrell Etherington, "Volkswagen to Offer Electric Versions of All of Its Vehicles by 2030," TechCrunch, https://techcrunch.com/2017/09/11 /volkswagen-to-offer-electric-versions-of-all-of-its-vehicles-by-2030/.

38. Peter Valdes-Dapena, "GM: The Future Is All-Electric," *CNN Money*, October 2, 2017, http://money.cnn.com/2017/10/02/technology/gm -electric-cars/index.html.

39. Truman Lewis, "Consumer Attitudes Towards Electric Cars Grow- ing More Positive, Survey Finds," *ConsumerAffairs*, September 19, 2016, https://www.consumeraffairs.com/news/consumer-attitudes-towards -electric-cars-growing-more-positive-survey-finds-091916.html.

40. "2018 Nissan LEAF S Specs," Nissan USA, last updated February 1, 2018, http://prods.nissanusa.com/electric-cars/leaf/versions-specs/version .s1.html.

41. Michael Sivak and Brandon Schoettle, "Relative Costs of Driving Electric and Gasoline Vehicles in the Individual U.S. States," The Uni- versity of Michigan, January 2018, http://umich.edu/~umtriswt/PDF /SWT-2018-1.pdf.

42. Leslie Shaffer, CNBC, "Electric Vehicles Will Soon Be Cheaper Than Regular Cars Because Maintenance Costs Are Lower, Says Tony Seba," June 14, 2016. https://www.cnbc.com/2016/06/14/electric-vehicles-will -soon-be-cheaper-than-regular-cars-because-maintenance-costs-are -lower-says-tony-seba.html.

43. Charles Fleming, "How Will I Charge My Electric Vehicle? And Where? And How Much Will It Cost?," *Los Angeles Times*, September 26, 2016, http://www.latimes.com/business/autos/la-fi-hy -agenda-ev-charging-20160920-snap-story.html.

44. "Plan to Create Electric Bus Fleet for Delhi," *Hindu*, November 19, 2017, http://www.thehindu.com/news/cities/Delhi/plan-to-create-electric -bus-fleet-for-delhi/article20554969.ece.

45. Nicolas Zart, "100% Electric Bus Fleet for Shenzhen (Population 11.9 Million) by End of 2017," CleanTechnica, November 12, 2017, https:// cleantechnica.com/2017/11/12/100-electric-bus-fleet-shenzhen-pop -11-9-million-end-2017/.

46. Judah Aber, "Electric Bus Analysis for New York City Transit," Columbia University, May 31, 2016, http://www.columbia.edu/~ja3041 /Electric%20Bus%20Analysis%20for%20NYC%20Transit%20by %20J%20Aber%20Columbia%20University%20-%20May%202016 .pdf.

47. MTA, "MTA Tests Electric Buses for Use on NY Streets," April 26, 2017, http://www.mta.info/news-nyct-bus/2017/04/26/mta-tests-electric -buses-use-ny-streets.

48. Bill Chappell, "Tesla Unveils Its Electric 'Semi' Truck, and Adds a Roadster," NPR, November 17, 2017, https://www.npr.org/sections /thetwo-way/2017/11/17/564777998/tesla-unveils-its-electric-semi-truck -and-adds-a-roadster.

49. Bernd Heid et al., "What's Sparking Electric-Vehicle Adoption in the Truck Industry?," McKinsey & Company, September 26, 2017, https:// www.mckinsey.com/industries/automotive-and-assembly/our-insights /whats-sparking-electric-vehicle-adoption-in-the-truck-industry.

50. Daimler, "The Mercedes-Benz Electric Truck," September 21, 2016, https://www.daimler.com/products/trucks/mercedes-benz/world -premiere-mercedes-benz-electric-truck.html.

51. Alex Crippen, "Warren Buffett Invests in Chinese Company Develop- ing 'Green' Cars," *CNBC*, September 27, 2008, https://www.cnbc.com /id/26916857.

52. Katie Fehrenbacher, "Electric Cars in China Are on Track for a Record Year," Greentech Media, October 20, 2017, https://www.greentechmedia .com/articles/read/electric-cars-in-china-are-on-track-for-a-record -year#gs.DBXTgho.

53. Sherisse Pham, "This Buffett-Backed Chinese Stock Is Up 55% in a Month," *CNN Money*, October 11, 2017, http://money.cnn.com/2017 /10/11/investing/byd-warren-buffett-china-electric-cars/index.html.

10. CONVERGENCE

1. California ISO, "Fast Facts," October 22, 2013, https://www.caiso.com /Documents/FlexibleResourcesHelpRenewables_FastFacts.pdf.

2. Sandia National Laboratories, "DOE Global Energy Storage Database," February 14, 2012, http://www.energystorageexchange.org/.

3. "Packing Some Power," *Economist*, March 3, 2012, http://www.economist .com/node/21548495?frsc=dg%7Ca.

4. *Lazard's Levelized Cost of Storage—Version 2.0*, December 15, 2016, https://www.lazard.com/media/438042/lazard-levelized-cost-of -storage-v20.pdf.

5. *Lazard's Levelized Cost of Storage*.

6. Tamra Johnson, "Americans Spend an Average of 17,600 Minutes Driving Each Year," AAA, September 8, 2016, http://newsroom.aaa .com/2016/09/americans-spend-average-17600-minutes-driving-year/.

7. Jess Shankleman, "Parked Electric Cars Earn $1,530 from Europe's Power Grids," *Bloomberg*, August 11, 2017, https://www.bloomberg.com /news/articles/2017-08-11/parked-electric-cars-earn-1-530-feeding-power -grids-in-europe.

8. Peter Campbell, "Electric Car Drivers to Sell Power Back to National Grid," *Financial Times*, May 10, 2016, https://www.ft.com/content /7e75b7d2-169c-11e6-b197-a4af20d5575e.

9. Stanley Reed, "Dutch Utility Bets Its Future on an Unusual Strategy: Selling Less Power," *New York Times*, August 18, 2017, https://www .nytimes.com/2017/08/18/business/energy-environment/eneco -netherlands-electricity-utility.html?_r=0.

10. Dom Galeon and Peter Caughill, "Soon, Tesla Cars Could Power the Grid (and Our Homes)," Futurism, last updated November 29, 2016, https:// futurism.com/soon-tesla-cars-could-power-the-grid-and-our-homes/.

11. James Ayre, "Vehicle-to-Grid Discharge, Even at Constant Power, Is Detrimental to EV Battery Performance, Study Finds," CleanTechnica, May 16, 2017, https://cleantechnica.com/2017/05/16/vehicle-grid-discharge -even-constant-power-detrimental-ev-battery-performance-study-finds/.

12. Kotub Uddin et al., "On the Possibility of Extending the Lifetime of Lithium-Ion Batteries Through Optimal V2G Facilitated by an Integrated Vehicle and Smart-Grid System," *Energy* 133 (August 2017): 710–722, http://www.sciencedirect.com/science/article/pii/S0360544217 306825?via%3Dihub#!.

13. National Renewable Energy Laboratory, "Connecting Electric Vehicles to the Grid for Greater Infrastructure Resilience," April 20, 2017, https://www.nrel.gov/news/program/2017/connecting-electric-vehicles -to-the-grid-for-greater-infrastructure-resilience.html.

14. Geoffrey Heal, "What Would It Take to Reduce U.S. Greenhouse Gas Emissions 80 Percent by 2050?," *Review of Environmental Economics and Policy* 11, no. 2 (Summer 2017): 319–335, https://geoffreyheal.files .wordpress.com/2017/08/reep-published.pdf.

15. Fred Lambert, "GM Announces Completed Production of 130 Autonomous Chevy Bolt EVs," Electrek, June 13, 2017, https://electrek.co /2017/06/13/gm-self-driving-chevy-bolt-ev/.

16. Anna Hirtenstein, "Move Over Tesla, Europe's Building Its Own Battery Gigafactories," *Bloomberg*, May 22, 2017, https://www.bloomberg .com/news/articles/2017-05-22/move-over-tesla-europe-s-building -its-own-battery-gigafactories.

17. Jason Deign, "10 Battery Gigafactories Are Now in the Works. And Elon Musk May Add 4 More," Greentech Media, June 29, 2017, https://www .greentechmedia.com/articles/read/10-battery-gigafactories-are-now -in-progress-and-musk-may-add-4-more?utm_source=Daily&utm _medium=Newsletter&utm_campaign=GTMDaily.

18. Peter Maloney, "California PUC Finalizes New 500 MW BTM Battery Storage Mandate," Utility Dive, May 4, 2017, https://www.utilitydive .com/news/california-puc-finalizes-new-500-mw-btm-battery -storage-mandate/441901/.

19. Adam Vaughan, "Is There Enough Electricity? National Grid Reacts to Fossil-Fuel Vehicle Ban," *Guardian*, July 26, 2017, https://www .theguardian.com/business/2017/jul/26/national-grid-fossil-fuel-vehicle -ban-electric-cars-is-there-enough-electricity-.

20. Tam Hunt, "Is There Enough Lithium to Maintain the Growth of the Lithium-Ion Battery Market?," Greentech Media, June 2, 2015, https:// www.greentechmedia.com/articles/read/is-there-enough-lithium-to -maintain-the-growth-of-the-lithium-ion-battery-m#gs.C3WeAfo.

21. Hunt, "Is There Enough Lithium?"
22. Amory Lovins, "Clean Energy and Rare Earths: Why Not to Worry," *Bulletin of the Atomic Scientists*, May 23, 2017, https://thebulletin.org /clean-energy-and-rare-earths-why-not-worry10785.
23. Richard Martin, "In Texas Oil Country, Wind Is Straining the Grid," *MIT Technology Review*, August 6, 2016, https://www.technologyreview .com/s/602112/in-texas-oil-country-wind-is-straining-the-grid/.
24. Benjamin Wehrmann, "The Energiewende's Booming Flagship Braces for Stormy Times," Clean Energy Wire, June 14, 2017, https://www .cleanenergywire.org/dossiers/onshore-wind-power-germany.
25. Jason Deign, "China Faces an Uphill Renewable Energy Curtailment Challenge," Greentech Media, November 17, 2017, https://www.green techmedia.com/articles/read/china-faces-uphill-renewable-energy -curtailment-challenge#gs.NM3xtKM.
26. "Gas Station Industry Statistics," StatisticsBrain, September 3, 2016, https://www.statisticbrain.com/gas-station-statistics/.
27. Fred Lambert, "US Has Now ~16,000 Public Electric Vehicle Charging Stations with ~43,000 Connectors," Electrek, June 19, 2017, https:// electrek.co/2017/06/19/us-electric-vehicle-charging-stations/.
28. Bloomberg New Energy Finance, *New Energy Outlook 2017*, June 15, 2017, https://about.bnef.com/new-energy-outlook/.
29. Bloomberg New Energy Finance, *Clean Energy Investment—3Q 2017 Trends*, October 5, 2017, https://about.bnef.com/blog/clean-energy -investment-3q-2017-trends/.
30. Giles Parkinson, "South Australia Already at 57% Wind and Solar in 2016/17," RenewEconomy, June 6, 2017, http://reneweconomy.com.au /south-australia-already-57-wind-solar-201617/.
31. Johnny Lieu, "Elon Musk Makes a Bet to Fix a State's Energy Woes in 100 Days, or It's Free," Mashable, March 10, 2017, http://mashable .com/2017/03/10/elon-musk-powerwall-australia/#ECTLg MPBV5qZ.

11. CONSEQUENCES

1. World Bank, "Access to Electricity (% of Population) 1990–2014," November 6, 2017, https://data.worldbank.org/indicator/EG.ELC .COAL.ZS?locations=CN-IN.

2. Beth Mole, "Delhi Becomes 'Gas Chamber' as Air Pollution Reaches Ludicrous Levels," Ars Technica, November 9, 2017, https://arstechnica.com/science/2017/11/off-the-charts-pollution-in-delhi-creates-gas-chamber-and-health-emergency/.

3. Bloomberg News with assistance by Feifei Shen, "Renewables Dominate China's New Capacity as Coal's Role Slips," Bloomberg, July 21, 2017, https://www.bloomberg.com/news/articles/2017-07-21/renewables-dominate-china-s-new-capacity-as-coal-s-role-slips.

4. SolarInsure, "Top 5 Largest Solar Power Plants of the World," last updated June 27, 2017, https://www.solarinsure.com/largest-solar-power-plants.

5. Andrew J. Stanley, Adam Sieminski, and Sarah Ladislaw, "China's Net Oil Import Problem," Center for Strategic & International Studies, April 10, 2017, https://www.csis.org/analysis/energy-fact-opinion-chinas-net-oil-import-problem.

6. Sanjeev Choudhary, "India's Dependence on Crude Oil Imports on Rise as Consumption Increases," Economic Times, April 22, 2016, https://economictimes.indiatimes.com/industry/energy/oil-gas/indias-dependence-on-crude-oil-imports-on-rise-as-consumption-increases/articleshow/51934359.cms.

7. Ed Crooks, "The Global Importance of China's Oil Imports," Financial Times, September 25, 2017, https://www.ft.com/content/e7d52260-a1e4-11e7-b797-b61809486fe2.

8. Paul Hockenos, "With Norway in Lead, Europe Set for Surge in Electric Vehicles," Yale Environment 360, February 6, 2017, http://e360.yale.edu/features/with-norway-in-the-lead-europe-set-for-breakout-on-electric-vehicles.

9. International Energy Agency, "Global EV Outlook 2017: Two Million and Counting," June 6, 2017, https://www.iea.org/publications/freepublications/publication/GlobalEVOutlook2017.pdf.

10. "Global Plug-in Sales for Q1-2018," http://www.ev-volumes.com/

11. "China to Build More Charging Points for Electric Vehicles," China Daily, February 10, 2017, http://www.chinadaily.com.cn/business/motoring/2017-02/10/content_28160372.htm.

12. Michael J. Coren, "China Is Selling More Electric Vehicles than the US—and It's Not Even Close," Quartz, May 3, 2017, https://qz.com/972897/china-is-selling-more-electric-vehicles-than-the-us-and-its-not-even-close/.

13. Jackie Wattles, "India to Sell Only Electric Cars by 2030, *CNN Money*, June 3, 2017, http://money.cnn.com/2017/06/03/technology/future/india -electric-cars/index.html.

14. "China Moves Towards Banning the Internal Combustion Engine," *Economist*, September 14, 2017, https://www.economist.com/news /business/21728980-its-government-developing-plan-phase-out -vehicles-powered-fossil-fuels-china-moves.

15. Charles Clover, "Subsidies Help China Sell the Most Electric Cars," *Financial Times*, October 23, 2017, https://www.ft.com/content/18afe28e -a1d2-11e7-8d56-98a09be71849.

16. "World's 10 Largest Auto Markets," *CNBC*, September 12, 2011, https://www.cnbc.com/2011/09/12/Worlds-10-Largest-Auto-Markets .html?page=11.

17. Phil LeBeau, "General Motors to Ramp Up Electric Vehicle Plans, 20 New Models Planned over Next 6 Years," *CNBC*, October 2, 2017, https://www.cnbc.com/2017/10/02/gm-to-ramp-up-electric-vehicle -plans-with-20-models-over-next-6-years.html.

18. Dana Varinsky, "Nearly Half of US Coal Is Produced by Companies that Have Declared Bankruptcy—and Trump Won't Fix That," *Business Insider*, December 9, 2016, http://www.businessinsider.com /us-coal-bankruptcy-trump-2016-12.

19. Kiran Stacey, "European Utilities Slash Asset Valuations," *Financial Times*, May 22, 2016, https://www.ft.com/content/5b2dd030-1e93-11e6-b286 -cddde55ca122.

20. "A World Turned Upside Down," *Economist*, February 25, 2017, https://www.economist.com/news/briefing/21717365-wind-and-solar -energy-are-disrupting-century-old-model-providing-electricity -what-will.

21. Guy Chazan, "RWE Posts €5.7bn Loss and Scraps Dividend," *Financial Times*, February 22, 2017, https://www.ft.com/content/4513da52-f8d0 -11e6-9516-2d969eod3b65?mhq5j=e2.

22. Anjli Raval and Andrew Ward, "Saudi Aramco Plans for a Life After Oil," *Financial Times*, December 10, 2017, https://www.ft.com/content /e46162ca-d9a6-11e7-a039-c64b1co9b482.

23. Clifford Krauss, "Norway's Wealth Fund Considers Divesting from Oil Shares," *New York Times*, November 16, 2017, https://www.nytimes .com/2017/11/16/business/energy-environment/norway-fund-oil.html.

24. Nadja Popovich, "Today's Energy Jobs Are in Solar, Not Coal," *New York Times*, April 25, 2017, https://www.nytimes.com/interactive/2017/04/25 /climate/todays-energy-jobs-are-in-solar-not-coal.html.

25. Environmental Defense Fund, "Now Hiring: The Growth of America's Clean Energy & Sustainability Jobs," January 24, 2017, http:// edfclimatecorps.org/sites/edfclimatecorps.org/files/the_growth_of _americas_clean_energy_and_sustainability_jobs.pdf.

26. Reuters Staff, "China to Plow $361 Billion into Renewable Fuel by 2020," *Reuters*, January 4, 2017, https://www.reuters.com/article/us -china-energy-renewables/china-to-plow-361-billion-into-renewable -fuel-by-2020-idUSKBN14P06P?platform=hootsuite.

27. International Renewable Energy Agency, "Turning to Renewables: Climate-Safe Energy Solutions," November 17, 2017, http://www.irena .org/-/media/Files/IRENA/Agency/Publication/2017/Nov/IRENA _Turning_to_renewables_2017.pdf.

28. Jennifer Granholm, "Transcript of Gov. Granholm's State of the State," *MLive*, February 3, 2009, http://www.mlive.com/politics/index .ssf/2009/02/live_video_gov_granholms_state.html.

29. Center for Health and the Global Environment, Harvard Medical School, *Mining Coal, Mounting Costs: The Life Cycle Consequences of Coal*, January 15, 2011, https://chge.hsph.harvard.edu/files/chge/files /MiningCoalMountingCosts.pdf.

30. Center for Health and the Global Environment, *Mining Coal, Mounting Costs*.

31. Paul R. Epstein et al., "Full Cost Accounting for the Life Cycle of Coal," *Ecological Economics Reviews* 1219, no. 1 (February 2011): 73–98, http://onlinelibrary.wiley.com/doi/10.1111/j.1749-6632.2010.05890.x /full.

32. Dana Loomis et al., "The Carcinogenicity of Outdoor Air Pollution," *Lancet Oncology* 14, no. 13 (December 2013): 1262–1263, http://www .thelancet.com/journals/lanonc/article/PIIS1470-2045(13)70487-X /fulltext.

33. Environmental Health and Engineering Inc., *Emissions of Hazardous Air Pollutants from Coal-Fired Power Plants*, Chicago State University Calumet Environmental Resource Center, March 7, 2011, https://www .csu.edu/cerc/researchreports/documents/EmissionsOfHazardous AirPollutantsFromCoal-FiredPowerPlants2011.pdf.

34. American Lung Association, "Health Risks of Particle Pollution," April 2017, http://www.lung.org/local-content/california/documents /particle-pollution-fact-sheet-2017.pdf.

35. World Bank and Institute for Health Metrics and Evaluation (IHME), *The Cost of Air Pollution: Strengthening the Economic Case for Action*, World Bank Group Open Knowledge Repository, 2016, https:// openknowledge.worldbank.org/bitstream/handle/10986/25013/108141 .pdf?sequence=4&isAllowed=y.

36. World Bank and IHME, *Cost of Air Pollution*.

37. Douglas W. Dockery et al., *Effect of Air Pollution Control on Mortality and Hospital Admissions in Ireland*, Health Effects Institute Report no. 176, July 27, 2013, https://www.healtheffects.org/system/files /Dockery-176.pdf.

38. Scripps Institution of Oceanography, "The Keeling Curve," last updated April 5, 2018, https://scripps.ucsd.edu/programs/keelingcurve/.

39. Intergovernmental Panel on Climate Change, "Summary for Policymakers," November 2, 2014, http://ar5-syr.ipcc.ch/topic_summary.php.

40. Josh Gabbatiss, "Worst-Case Global Warming Predictions Are the Most Accurate, Say Climate Experts," *Independent*, December 6, 2017, http://www.independent.co.uk/environment/global-warming -temperature-rise-climate-change-end-century-science-a8095591 .html.

41. International Energy Agency, *Energy and Climate Change: World Energy Outlook Special Report*, June 15, 2015, https://www.iea.org/publications /freepublications/publication/WEO2015SpecialReportonEnergyand ClimateChange.pdf.

42. International Renewable Energy Agency, "Turning to Renewables."

43. International Renewable Energy Agency, "Turning to Renewables."

44. International Renewable Energy Agency, "Turning to Renewables."

45. Peter Applebome, "They Used to Say Whale Oil Was Indispensable, Too," *New York Times*, August 3, 2008, http://www.nytimes .com/2008/08/03/nyregion/03towns.html.

46. Derek Thompson, "The Spectacular Rise and Fall of U.S. Whaling: An Innovation Story," *Atlantic*, February 22, 2012, https://www.theatlantic .com/business/archive/2012/02/the-spectacular-rise-and-fall-of-us -whaling-an-innovation-story/253355/.

47. Thompson, "Spectacular Rise and Fall."

48. New Bedford Whaling Museum, "Timeline: 1602 to Present," November 7, 2013, https://www.whalingmuseum.org/learn/research -topics/timeline-1602-to-present.

49. Applebome, "Whale Oil."

50. William Finnegan, "Is Donald Trump Already Forsaking Coal Country?," *New Yorker*, July 18, 2017, https://www.newyorker.com/news /daily-comment/is-donald-trump-already-forsaking-coal-country.

51. Taylor Kuykendall and Ashleigh Cotting, "Companies Recently Filing Bankruptcy Produce More than 2/3 of PRB Coal," SNL Interactive, April 13, 2016, https://www.snl.com/InteractiveX/Article .aspx?cdid=A-36118340-12086.

52. Javier Blas, "Remember Peak Oil? Demand May Top Out Before Supply Does," *Bloomberg Businessweek*, July 11, 2017, https://www .bloomberg.com/news/articles/2017-07-11/remember-peak-oil -demand-may-top-out-before-supply-does.

53. BP, *BP Energy Outlook: 2018 Edition*, February 20, 2018, https://www .bp.com/content/dam/bp/en/corporate/pdf/energy-economics /energy-outlook/bp-energy-outlook-2018.pdf.

54. Institute for Energy Economics and Financial Analysis, "IEEFA Report: Winners and Losers Among Big Utilities as Renewables Disrupt Markets Across Asia, Europe, the U.S., and Africa," October 4, 2017, http://ieefa.org/ieefa-report-winners-losers-global-electricity-market -renewables-disrupt-markets-across-asia-europe-u-s-africa/.

55. Institute for Energy Economics and Financial Analysis, "IEEFA Report."

56. Pilita Clark, "Kingdom Built on Oil Foresees Fossil Fuel Phase-Out This Century," *Financial Times*, May 21, 2015, https://www.ft.com /content/89260b8a-ffd4-11e4-bc30-00144feabdco.

57. Rania El Gamal, Reem Shamseddine, and Katie Paul, "Saudi Arabia Pushes Ahead with Renewable Drive to Diversify Energy Mix," *Reuters*, April 17, 2017, https://www.reuters.com/article/saudi-renewable/saudi -arabia-pushes-ahead-with-renewable-drive-to-diversify-energy-mix -idUSL8N1HP1oB.

12. NO TIME TO LOSE

1. Carbon Brief, "Analysis: Global CO2 Emissions Set to Rise 2% in 2017 After Three-Year Plateau," https://www.carbonbrief.org/analysis-global-co2 -emissions-set-to-rise-2-percent-in-2017-following-three-year-plateau.

2. Scripps Institution of Oceanography, "The Keeling Curve," last updated June 28, 2018, https://scripps.ucsd.edu/programs/keelingcurve/.

3. Intergovernmental Panel on Climate Change, "Summary for Policymakers," November 2, 2014, http://ar5-syr.ipcc.ch/topic_summary.php.

4. Eric Morris, "From Horse Power to Horsepower," *Access* 30 (Spring 2007): 2–9.

5. Jack Perkowski, "China and the U.S. Supercharge the Growing Global Electric Vehicle Industry," *Forbes*, February 28, 2017, https://www.forbes.com/sites/jackperkowski/2017/02/28/china-and-the-u-s-supercharge-the-growing-global-electric-vehicle-industry/#288da92b2454.

6. Nikolas Soulopoulos, "When Will Electric Vehicles Be Cheaper than Conventional Vehicles?," Bloomberg New Energy Finance, http://www.automotivebusiness.com.br/abinteligencia/pdf/EV-Price-Parity-Report.pdf.

7. "Chevy Bolt Battery Cell Cost," Inside EVs, accessed January 10, 2018, https://insideevs.com/wp-content/uploads/2015/10/bolt-battery-cost-lg-chem.jpg.

8. Soulopoulos, "Electric Vehicles."

9. Henry Fountain and Derek Watkins, "As Greenland Melts, Where's the Water Going?," *New York Times*, December 5, 2017, https://www.nytimes.com/interactive/2017/12/05/climate/greenland-ice-melting.html?_r=0.

10. Richard Z. Poore, Richard S. Williams Jr., and Christopher Tracey, "Sea Level and Climate," U.S. Geological Survey, last modified November 29, 2016, https://pubs.usgs.gov/fs/fs2-00/.

11. Poore, Williams, and Tracey, "Sea Level and Climate."

12. Joby Warrick and Chris Mooney, "Effects of Climate Change 'Irreversible,' U.N. Panel Warns in Report," *Washington Post*, November 2, 2014, https://www.washingtonpost.com/national/health-science/effects-of-climate-change-irreversible-un-panel-warns-in-report/2014/11/01/2d49aeec-6142-11e4-8b9e-2ccdac31a031_story.html?utm_term=.41a0c1bb1cff.

13. Stephane Hallegatte et al., "Future Flood Losses in Major Coastal Cities," *Nature Climate Change* 3 (August 2013): 802–806, https://www.nature.com/articles/nclimate1979.

14. Warrick and Mooney, "Effects of Climate Change."

INDEX

Page numbers in *italics* indicate figures

Abu Dhabi: solar power in, 106
acid rain, 61–62, 100
Africa: pay-as-you-go model in, 69
air pollution. *See* pollution
Anglo-Persian Oil Company, 78
Antarctica, 157–158
Appalachia, 140. *See also* coal
Apple, 104
Aramco, 138
Asia: smog in, 112. *See also* China;
   India
asthma, 12
atomic energy. *See* nuclear power
auctions, 57–58, 105–106. *See also*
   government incentives and
   subsidies
Australia: solar panels in, 103, 129
automobiles. *See* cars; electric
   vehicles
autonomous electric vehicles,
   123–124

batteries: and climate change,
   156–157; degradation of, 122–123;
   disadvantages of, 82; efficiency
   of, 120–121; energy density of, 82,
   88; and Tesla, 84–89, 91–94, 124,
   129. *See also* BYD; Gigafactory;
   range anxiety
behind-the-meter energy, 49, 72
Bell Labs, 43
Benz, Carl, 76, 156
Benz Patent-Motorwagen, *74*
Berkshire Hathaway: investment
   in BYD, 113–114; investment in
   solar and wind, 107–108
biofuels, 24–25
biomass energy, 23–25
black lung disease, 139
BlackRock, 108
blockchain, 123
Bloomberg New Energy Finance,
   157

BMW, 122

BP, 145, 146

Brazil: biofuel production of, 25; Earth Summit, 59; hydropower projects in, 22; Rio de Janeiro, 54

Brexit, 148

Brimblecombe, Peter, 9

Britain. See Great Britain

Buffett, Warren, 96; investment in BYD, 113–114; investment in solar and wind, 97, 107–108

buses: electric, 112–114

Bush, George H. W., 61

BYD, 113–114, 135

California: cap and trade in, 63; duck curve of, 117–119; energy-storage law, 124; and grid parity, 98; load profile of, 118–119, 119; and solar panels, 103; and zero-emission vehicles, 83, 94

cap and trade. See carbon markets

capacity factor: defined, 32–33; of wind, 38

Cape Wind, 38–39

capital markets, 106–107

carbon dioxide: concentrations of 155; emissions of, 152, 152–154, 158; greenhouse effect caused by, 141–142; pollution caused by, 100

carbon markets, 59–63

carbon sequestration, 155–156

cars: Benz Patent Motorwagen, 74; introduction of, 76–77; replacing horses, 2, 76, 156; self-driving, 123–124. See also electric vehicles

centralized generation, 48

charge stations, 86, 94–95, 127, 135

Chevy Bolt, 111, 124

Chile: solar power in, 105

China: carbon markets in, 63; coal usage of, 134; electric vehicles in, 87, 109–110, 112–114, 134–136; jobs in renewable energy sector in, 138–139; and Kyoto Protocol, 62–63; oil consumption of, 134–135; president of, 132, 133; Three Gorges Dam, 20, 22, 23; transition to renewable energy of, 134–136; wind power in, 21, 63, 127, 134

Churchill, Winston, 78

Clean Air Act (1990), 61–62

climate change, 100, 141–142, 148, 151–159

Climate Change Convention, 59

coal: benefits of, 8; compared to gas, 76; compared to whaling industry, 145; decline of, 136–137; drawbacks of, 9; employment in, 145; as percentage of U.S. energy source throughout history, 18; pollution from, 12, 61–62, 100, 139–140; price of, 9, 11; public health costs of, 140; and Royal Navy, 77–78; transition to nuclear power from, 13–15; transition to gas from, 15–17; transition from wood to, 8–11; U.K. consumption of, 9–11; U.S. consumption of, 9–11; use in China and India, 134

coal mines: deaths from, 139; depletion of, 11

Columbia Business School, viii

Columbia University, 112, 157

combined-cycle gas turbines, 16

command and control incentives, 56–57

compressionless engine, 76

concentrated solar power (CSP), 50

Consumer Federation of America, 111

contagion effect, 103

convergence, 117–130

conversion efficiency, 45

corn: used for biofuel, 24

Cronkite, Walter, 39

CSP. *See* concentrated solar power

da Vinci, Leonardo, 76

Daimler, 124

Dales, John, 61

dams, 20, 22, 23. *See also* hydropower

deforestation, 8

degradation, 122–123

Denmark: offshore wind farm near, 106; taxes on electric vehicles in, 88; vehicle-to-grid storage trial in, 121–122

deregulation, 16

developing nations: benefits of renewable energy for, 147; feed-in tariffs in, 56; off-grid solar projects in, 48; pay-as-you-go model in, 69–70; wood burning in, 23

diesel, 24, 112–113, 135, 138

distributed energy, 48, 72, 102

Domesday Book, 7

Dublin: coal ban in, 140

duck curve, 117–119

Earth Summit (1992), 59

Edison, Thomas, 11, 48, 72; partnership with Henry Ford, 81–82

Einstein, Albert, *42*, 43

Eisenhower, Dwight D., 14

electric grid: and load profile, 118; and renewable energy, 72; and storage, 121–122. *See also* electricity

electric vehicles, 81–95; in China, 87, 109–110, 112–114, 134–136; and climate change, 156–157; compared to internal combustion engines, 81–82, 109; costs of, 89–90, 111; financing for, 128; government incentives for, 87–88, 135; and grid parity, 108–113; history of, 81–83; in Norway, 138; reduction of carbon dioxide emissions, 152; sales of, *110*; self-driving, 123–124; solar-powered, *116*; trucks and buses, 112–114; in the U.S., 81, 109–110; vehicle-to-grid storage using, 121–124. *See also* batteries; Tesla

electricity: challenges to sector, 72, 136–137; demand for, 117–119, 125; economic impact of, 12–13, 17–18; generated by coal, 10–11; generated by gas, 16; generated by nuclear power, 14–15; generated by solar, 45, 50, 53; generated by wind, 32–34, 40; government regulation of, 128; storage of, 119–122; U.S. consumption of, 122. *See also* grid parity; LCOE; load profile; utilities; vehicle-to-grid storage

Emerson, Ralph Waldo, 8
employment, 138–139
Enel utility, 121–122, 146
energy crisis (1973), 29, 44, 83
energy decarbonization, 142–143
energy density: of batteries, 82, *88*;
    of coal, 8; defined, 8; of waves, 23
Energy Information
    Administration, 17, 18
energy transitions: and carbon
    markets, 61; climate
    consequences of, 141–143,
    152–154; economic consequences
    of, 136–139; from coal to
    nuclear, 13–15; from fossil fuels
    to renewables, 21–27, 134–136;
    from horses to cars, 2, 76, 156;
    geopolitical consequences of,
    133–136; health consequences
    of, 139–140; hurdles to, 71–72;
    from oats to oil, 75–79; risks
    and benefits of, 3–4; from wood
    to coal, 8–11. *See also* electric
    vehicles; renewable energy;
    solar power; wind power
England: population growth of, 8;
    and project finance, 67; wind
    farms in, 67, 106. *See also* Great
    Britain; United Kingdom
environmental impact. *See* climate
    change
environmental legislation, 12. *See
    also* carbon markets; government
    incentives and subsidies
E.ON utility, 137
ethanol, 24–25

EU Emissions Trading System, 63
Europe: and electric vehicle parity,
    109–110; government incentives
    for renewable energy, 33–34, 63;
    transition to renewable energy,
    148; utility struggles of, 136, 146;
    wind farms in, 40
European Wind Energy
    Association, 30
EV1, 83–84
externalities, 100

feed-in tariffs, 56–57
Fields, Mark, 95
Ford, Henry: partnership with
    Thomas Edison, 81–82
Ford Model T, 76
Ford Motor Company, 95
forests: burning of, 7–8
fossil fuels: bans on, 135; and
    climate change, 142–143;
    compared to whaling industry,
    144–145; divestment from,
    138; government subsidies for,
    99–100; pollution from, 12,
    61–62, 100, 139–140; transition to
    renewables from, 21–27, 134–136.
    See also coal; gas; petroleum
fracking, 16–17
France: ban on fossil fuel–powered
    vehicles in, 135; and project
    finance, 66–67; tidal and wave
    power in, 23; wind power in,
    66–67
Franklin, Benjamin, 10
Füger, Heinrich, 6

gas: compared to coal, 12, 17, 76; consumption of, 77; economics of, 15–17; as percentage of U.S. energy source throughout history, *18*; transition from coal to, 15–17; transition to electric from, 138; usage in China and India, 134

gearless turbines, 33, 102

General Motors: battery cost goal, 157; electric vehicles, 83–84, 111, 124, 136

geothermal power, 25–26

Germany: Gigafactory in, 124; transmission lines for wind power in, 127; use of auctions for renewable energy, 58; utilities in, 137, 146

Gigafactory, 91–92, 124, 126, 129

global energy crisis (1973), 29, 44, 83

global warming, 141–142, 151, 153, 157–158. *See also* climate change

GM. *See* General Motors

Godalming, England, 10

Goldman Sachs, 108

government incentives and subsidies, 56–64; in China, 135; for electric vehicles, 87–88; feed-in tariffs, 56–57; for fossil fuels, 99–100; and grid parity, 99–100; market mechanisms, 57–63; for wind power, 33–34

government regulations: as barriers to renewable energy, 128; and electric grid, 72–73; and energy storage, 124

Granholm, Jennifer, 139

Great Britain: ban on fossil fuel–powered vehicles in, 135; coal usage of, 2, 77–78; depletion of coal mines in, 11; first electric power station in, 10; transition to coal, 8–11; transition of Royal Navy fleet to oil, 77–78; wood-burning of, 8. *See also* England; United Kingdom

Great Smog of London (1952), 12

greenhouse gas: and California, 63; and climate change, 141–142, 151–159; and Climate Change Convention, 59; flows, 151–153; and Kyoto Protocol, 59–63; reduction through carbon markets, 61–62; stock, 153, 158

Greenland, 157

grid. *See* electric grid

grid parity, 97–114; defined, 97; and electric vehicles, 108–113; and government subsidies, 99–100

Hansen, James, 151, 157

Harvard Medical School, 139

Hawaii, 141

health consequences, 12, 139–140. *See also* pollution

Hoover Dam, 22

horses, 2, 75–76, 156

Hurricane Katrina, *150*

hybrid vehicles, 87, 92, 111. *See also* electric vehicles

hydraulic fracturing. *See* fracking

hydropower: limitations of, 22–23; as percentage of U.S. energy source throughout history, *18. See also* pumped hydro storage

ice sheets: melting of, 157–158
India: electric buses in, 112; solar power in, 105; transition to renewable energy, 134–136; wind power in, 106
Industrial Revolution: and carbon dioxide, 142; and coal, 9; and hydropower, 22
installed capacity of wind turbines, 35
Institute for Energy Economics, 145
Institute for Health Metrics and Evaluation, 140
intermittency: in California, 118; electric vehicles as solution to, 123, 125, 129, 130; of wind and solar, 38, 73, 100–101, 117–118
internal combustion engine: compared to electric vehicles, 81–82, 109, 111; energy transition from, 2–3; history of, 76
International Energy Agency, 99
International Monetary Fund, 1
International Renewable Energy Agency (IRENA), 139, 142, 152, 155
Ireland: coal ban in, 140
Italy: and Enel utility, 121–122, 146

Japan: and Kyoto Protocol, 59; sales of electric vehicles in, 110; and Tepco utility, 146

Kansas: wind farm in, 28
Keeling, Charles, 141
Keeling Curve, *141*, 141–142, 153
Kenya: geothermal power of, 26
Korea. *See* South Korea
Kyoto Protocol, 59, 61–63

La Cour, Poul, 29
Lazard, 15, 66, 120
LCOE: and cost of capital, 66; defined, 13; formula, 161–162; of gas, 17; of geothermal, 26; of solar, 45–46, 48–53, 66, 71, 101; and utility "death spiral," 137; of wind, 33–36, *36*, 66, 71
Leaf (Nissan), 93, 111
learning curve: applied to batteries, 89, 92; applied to nuclear power, 14–15; applied to solar power, 48–53, 71; applied to wind power, 33–36, 71; defined, 34–35
levelized cost of electricity. *See* LCOE
light bulb: invention of, 10
liquid fuel. *See* diesel; ethanol; gas; petroleum
lithium, 126
lithium-ion batteries: degradation of, 122–123; efficiency of, 120–121; and energy transitions, 164; manufacturing cost of, 156–157; and Nissan, 93; and Tesla, 85, 88–89, 91, 124. *See also* BYD; Gigafactory
load profile, 117–118
London: and horses, 75; in seventeenth century, 8; and smog, 12

London Array, 67

lumber. *See* wood

market mechanisms, 57–63

Massachusetts whaling industry, 144

Mastercard, 69

Mauna Loa, 141

McKinsey & Company, 113

Mercedes-Benz, 113, 124

mercury, 12, 100

Michigan, 139

Mitchell, George, 17

M-KOPA Solar, 69

Model S (Tesla), 86–87; running costs of, 89–91; range of, 111

Model T (Ford), 76

Model 3 (Tesla), 92

Model X (Tesla), 111

Moore, Gordon, 35

Moore's law, 35

Musk, Elon, 80; and batteries, 88–89, 91; and "Secret Tesla Motors Master Plan," 84; and South Australia electricity storage, 129. *See also* Gigafactory; Tesla

al-Naimi, Ali, 147

NASA, 44, 46

National Grid (U.K.), 125

natural gas. *See* gas

net metering: defined, 49; disputes over, 72; prohibited, 128

Netherlands: government incentives for electric vehicles, 87–88; vehicle-to-grid storage trial in, 122

Nevada, 91

New Bedford, Massachusetts, whaling industry, 144

New Delhi, 112

New Orleans, 150

New York City: consequences of climate change to, 157–158; electric buses in, 112–113; electric taxis in, 81; first power plant in, 11; and horses, 75

NextEra utility, 146

NIMBYism, 38–39, 126

Nissan, 93, 111, 121–122

Norwegian Government Pension Fund, 138

NRG utility, 146

nuclear power: economics of, 13–15; first reactor, 14; LCOE of, 14; as percentage of U.S. energy source throughout history, 18; transition from coal to, 13–15

oats: horse consumption of, 75

oil: Chinese production of, 134; as percentage of U.S. energy source throughout history, 18; impact of electric vehicles on, 138; and U.S. transportation, 79; well, 77; from whales, 143–144. *See also* energy crisis

ore refining, 8

Oregon State University, 23

Oxford Institute for Energy Studies, 137

parity. *See* grid parity

particulates, 113, 134, 140

pay-as-you-go (PAYG), 69–70

peak load, 118
Pearl Street Station, 11, 48, 72
Pennsylvania: discovery of oil in,
    144; Three Mile Island, 14
Persians: inventors of windmills,
    29
petroleum: replacement by biofuels,
    24; U.S. production of, 144; U.S.
    usage of, 78. *See also* fossil fuels;
    gas; oil
photovoltaic solar cells: invention of,
    43. *See also* solar panels
pipelines: history of, 16
plant biofuels, 24–25
plant biomass, 7, 23–24
pollution: from coal, 12, 61–62, 100,
    139–140; in Asia, 112, 134; from
    particulates, 113, 134, 140; from
    nuclear power, 13–14; reduced by
    carbon markets, 61–63; reduced
    by electric vehicles, 112–113
population growth: in Asia, 16; in
    Europe, 8
Porsche, Ferdinand, 81
power plants: economics of, 12; first,
    10–11; coal vs. gas, 12; CSP, 50;
    gas-fired, 17; geothermal, 26;
    renewable energy predictions for,
    21; as source of pollution, 140
power purchase agreement (PPA),
    65–66
Prius (Toyota), 92
project finance, 64–67
Prometheus Brings Fire to Mankind, 6
prosumers, 103–104
pumped hydro storage, 119–120
PV. *See* solar panels

radioactive waste: disposal of, 14
Ramana, M. V., 15
range anxiety, 86, 94–95
rare earth metals, 126
regulations: as barriers to renewable
    energy, 128; and electric grid,
    72–73; and energy storage, 124
Renault, 93. *See also* Nissan
renewable energy: and climate
    change, 142–143, 154; corporate
    transition to, 104; corporate
    investment in, 108; defined,
    21–22; employment in, 138–139;
    financing of, 55–73; government
    incentives for, 56–63, 135;
    misconceptions about cost of,
    vii–viii; power plants, 21. *See also*
    electric vehicles; hydropower;
    solar power; wind power
renewable portfolio standard. *See*
    RPS markets
Rio de Janeiro, 54, 59
Ritter, Bill, vii–viii
Rive, Lyndon, 98
Roadster (Tesla), 85–86
Royal Dutch Shell, 145
Royal Navy (British), 77–78
RPS markets, 58–59, 60
RWE utility, 137, 146

satellites: solar-powered, 44
Saudi Arabia: investment in
    renewables of, 147; and oil, 138;
    solar power of, 2
self-driving cars, 123–124
Shenzhen, China, 112, 114
silicon, 126

Simon, Bill, 104

smart charging, 123

smelting, 8

smog: in Asia, 112; in London, 12

Smoky Hills Wind Farm, *28*

solar leasing, 68–70, 97–98

solar panels: auction prices of, *58*; and contagion effect, 103; and energy transitions, 164; manufacturing and operating costs of, 45–53; number installed per day, 21; permits for, 127; raw materials used for, 126; in the U.S., 50, 103; use in space, 44, 46; vehicle-to-grid storage using, 122–123. *See also* solar leasing; solar power

solar power, 43–53; in Abu Dhabi, 106; annual growth of market, 52; in Chile, 105; in China, 134; commercial projects, 49; compared to biofuel, 25; corporate investment in, 108; CSP projects, 50; as distributed energy, 102; economics of, 44–53; and electric vehicles, *116*, 122–123; financing for, 128; first application of, 44; in India, 105, 134; institutional investment in, 70–71; intermittency of, 73, 100–101, 117–118; job creation of, 138–139; LCOE of, 45–46, 48–53, 66, 71, 101; residential projects, 48–49, 68–70, 98, 102–103; in Saudi Arabia, 2, 147; trackers for, 100–101; in the U.S., 69, 105, 138; utility-scale projects, 49–50, *52*, 65–66. *See also* solar leasing; solar panels

SolarCity, 98, 122

Sony, 85

soot, 12

South Australian electricity storage, 129

South Korea: tidal and wave power of, 23

Spindletop, Texas: oil well, *77*, 144

steam engine: invention of, 9; and power plant, 11

storage. *See* batteries; vehicle-to-grid storage

subsidies. *See* government incentives and subsidies

sugarcane, 25

sulfur dioxide, 61–62, 100

SunPower Corporation, 47, 116

superchargers, 86

Surrey, England, 10

Swanson, Richard, 47

Swanson's law, *47*, 48

Sweden: wood usage of, 23–24

tax incentives. *See* government incentives and subsidies

Tepco utility, 146

Tesla, 80, 84–94; and batteries, 84–89, 91–94, 124, 129; direct-distribution model of, 90; formation of, 84; range of cars, 111; and South Australia electricity storage, 129; trucks, 112–113; vehicle-to-grid storage, 122. *See also* Gigafactory; Model S; Model X; Musk, Elon; Roadster

Tesla, Nikola, 84

Texas: oil well, 77; wind power of, 2, 127

Three Gorges Dam, 20, 22, 23

Three Mile Island, 14

tidal power, 23

timber. *See* wood

Titusville, Pennsylvania, 144

Toyota, 92

trackers, 101–102

transmission lines, 39, 123, 127, 128

transportation sector: transition to electric vehicles, 2–3, 112–114

trucks: electric, 112–114

turbines: gas, 16; wind, 21, 29–32, 35, 37, 102

UBS, 109–110, 117

United Kingdom: consumption of coal, 9–11; demand for electricity in, 125; vehicle-to-grid storage trial, 121–122. *See also* England; Great Britain

United States: air pollution in, 12; and biofuels, 24; charging stations in, 127; and Clean Air Act, 61–62; and Climate Change Convention, 59; coal usage of, 9, 136; consequences of climate change to, 157–158; and electric vehicles, 81, 109–110; electricity consumption of, 122; energy sources throughout history, *18*; fossil fuel subsidies, 99; government policy of, 148; hydropower of, 22; mining deaths in, 139; and net metering, 128; nuclear power of, 14–15;

pipeline network of, 16; and RPS markets, 58–60; and sales of Model S, 90; solar panel installations of, 50, 103; solar sector growth in, 69, 105, 138; states at grid parity, 98; tax credits for renewable energy, 57, 87; transition to coal, 9–11; transition to gas, 16–17; transition of U.S. Navy fleet to oil, 78–79; utilities, 146; whaling industry, 143–144; wind farms in, 40; wind power in, 105–106; windmill usage of, 29; wood usage of, 8–9. *See also* California; New York City; Texas

uranium, 12–13

U.S. Department of Energy, 123

U.S. Energy Information Administration, 17, 18

U.S. Navy, 78–79

utilities: consumer distrust of, 102; regulation of, 72–73, 128; struggles of, 72, 136–137, 145–146; transition to renewable energy of, 105–106. *See also* net metering

utility-scale projects, 49–50, 52; financing of, 65–67; solar and wind, 105

van Beurden, Ben, 145

Vanguard I satellite, 44

vehicle-to-grid (V2G) storage, 121–124

Volkswagen, 111

Volvo, 111

Wales: population growth of, 8
Walmart, 104
Wang Chuanfu, 113, 135
water. *See* hydropower
Watt, James, 9
wave power, 23
Wesleyan University, 158
whaling industry, 143–145
Who Killed the Electric Car?, 83
William the Conqueror, 7
wind farms: capacity factor of, 38;
    London Array, 67; offshore, 106;
    operational costs of, 31–34, 37;
    Smoky Hills Wind Farm, 28
wind power, 29–40; in China, 21, 63,
    127, 134; corporate investment
    in, 108; economics of, 31–37;
    financing of, 128; history of, 29;
    importance of wind speed to, 30,
    38; in India, 106; innovations in,
    102; institutional investment in,
    70–71; intermittency of, 38, 73,
    100–101, 117–118; job creation of,
    138–139; LCOE of, 33–36, 66, 71;
    subsidies for, 33–34; transmission
    lines for, 127; in the U.S.,

105–106; utility-scale projects,
    65–67.

wind turbines: in China, 21;
    construction and installation
    of, 29–32, 37; in Denmark, 29;
    and energy transitions, 164;
    innovations in, 102; installed
    capacity of, 35
windmills, 29
wood: burning of, 7; energy density
    compared to fossil fuels, 24; as
    percentage of U.S. energy source
    throughout history, 18; price of, 8
World Bank, 140, 158
World Health Organization, 112,
    134, 140
World Resources Institute, 24–25

Xi Jinping, 132, 133

Yohe, Gary, 158

zero-emission vehicles: and
    California, 83; trucks and buses,
    112–113. *See also* electric vehicles

CPSIA information can be obtained
at www.ICGtesting.com
Printed in the USA
LVHW090038110419
613744LV00001B/1/P

9 780231 187848